TOWARD an ATLAS
of the European Delta Landscape

Maria Chiara Tosi

Foreword — 6

Between Fragility and Opportunity — 8

Part One
European Delta Landscape

Biography and comparison as instruments of knowledge — 14

Geographies and comparison — 18
Biographies
◇◇◇ Danube ◇◇◇ — 54
◇◇◇ Ebro ◇◇◇ — 62
◇◇◇ Guadalquivir ◇◇◇ — 70
◇◇◇ Nemunas ◇◇◇ — 78
◇◇◇ Po ◇◇◇ — 86
◇◇◇ Rhine ◇◇◇ — 94
◇◇◇ Rhone ◇◇◇ — 102

Part Two
Po Delta

Some Ipothesis — 110

Explorations in the Po Delta — 114

Notes on the visual identity of the territory — 116
Po Delta 2100 — 127
Actions for a deltaic resilience — 138

Foreword

This book addresses the deltaic territories in Europe that are both fragile and affected by important climate changes.

These characteristics, one inherent to the regions under study, the other caused by their transformation over time, are producing significant effects on these places and on the ways they are inhabited. The book is organised in two main parts.

The first part proposes an analysis of the most important features of some of the European Deltas, and compares them, in order to identify their common as well as specific and particular problems. The Atlas refers to the respective deltas of the Danube, the Ebro, the Guadalquivir, the Nemunas, the Po, the Rhine and the Rhone. The analytical criteria taken into consideration, being the most important criteria used, makes special reference to the ecological problems affecting these territories (mainly linked to dangers from the water and from climate change), but it also examines the specific social and economic impacts deriving from these problems.

Biography and the comparison of maps are the tools used to better understand these deltaic territories. What will the Po river delta region look like in 2100? Envisioning such a scenario is the key endeavour of the second part of this volume. As with many other deltaic territories, the Po delta is a fragile region, posing both social paradoxes and ecological dilemmas. This part argues that today, at the beginning of the 21st century, it is of the utmost importance to envision the fate of similar territories, affected by the drastic phenomena of ongoing climate change.

This part aims at dealing with both the critical conditions and the potential resources of the Po Delta region, in order to define new visions for the territory and to try to overcome its current weakness. The book intends to highlight the necessity of reflecting on the delta landscapes in Europe, which represent a particular case of the deltaic territories in the world. A comparison of cases among different European countries with different characteristics, threats, traditions and policies thus addresses these issues on the European scale, and with the intention of to placing the deltaic territories back at the centre of local and European discussion.

This book is the result of several years of research and design experiments conducted as part of the Observatory on the landscape of the Po Delta and during an Erasmus Intensive Programme founded by EC which saw the participation of several European universities: the Ion Mincu University of Bucharest, Vilnius Gedminas Tecnichal University, the Technical University of Delft, the University of Sheffield, the Autonomous University of Barcelona, the Escola Superior Technical del Valles and finally the Iuav University of Venice. I thank all those colleagues and students who with their work and reflections have contributed to advancing the knowledge about and projects for the European deltas.

Thanks also go to those who contributed to study by furnishing information and knowledge: Alexandru Drosencu of the Danube Delta National Research and Development Institute, Nerijus Gricevicius of the Nemunas Delta Regional Park, Inmaculada Juan Franch of the Ebro Delta Natural Park, Francisco Quirós Herruzo of the Donana Natural Park, and Sara Ariano, Francesco Baratti, Marina Bertoncin, Davide Fornari, Massimo Mazzanti, Armand Mevis, Luciano Perondi and Sybrand Tjallingii.

I also wish to thank the Consortium of Reclamation for the Po Adige Delta, and in particular Lino Tosini and Giancarlo Mantovani for their valuable advice and the opportunity that their extensive knowledge of the area offered to test hypotheses and ideas.

Finally, a big thank you goes to the local community of the Po Delta, which hosted our initiatives and made us see the territory in a way that would otherwise have been impossible.

Between Fragility
& Opportunity

The delta areas in the world are dynamic systems where considerable natural resources interact with high population density and significant production potential. These areas of confluence between fresh and salt water are among the most active and innovative areas on the planet, where fast-growing metropolises compete with wetlands with a high level of biodiversity: however, the conflicts between the needs of the land and those of the water that for centuries have continually redesigned the geographies of these territories and the civilizations that inhabit them, in recent decades have dramatically reduced the resilience of these areas.

The climate changes affecting the entire planet are radicalized in these regions, so that coastal erosion, the rise in the average sea level, subsidence, the intrusion of sea water with the subsequent salinisation of groundwater and irrigation systems, flooding and droughts all reveal the increasing fragility of the deltas, jeopardizing the very survival of the people who live there and the productive activities located there.

Irreversible damage to local infrastructure severely undermines the quality of life in these areas which demonstrate a clear need for integrated interventions where the management of environmental risks and the design of quality living spaces converge in unified strategies.

These issues are the focus of a recent international debate that through dialogue between similar territories aims to examine the effectiveness of various strategies and methods of intervention. (www.delta-alliance.nl/deltas) The comparison between expert knowledge and local knowledge on the one hand and between territories that while differing in environmental, settlement and socio-economic features show similarities in the fragilities and in the actions needed to overcome them on the other, is the hallmark of the ways in which innovations in knowledge about the deltas are being sought. (Bucx et al. 2010)

Within this debate, the European deltas occupy a special place because despite being affected by the same type of fragilities, they are characterized by far different settlement conditions, and this has led to exploring innovative intervention strategies. Unlike other deltas areas, with the exception of the Rhine Delta, which is configured as a highly urbanized area, we are dealing in fact with areas of low population density, mainly farmlands that have long been considered by the national and regional territorial policies as empty space, to fill with functions rejected elsewhere (nuclear power plants, power plants, regasification plants, alternative energies, etc.).

If on the one hand the dynamic nature of the relationships between land and water, the constant change that affects the whole system of operation of these territories has facilitated the formation of prized environments, habitats of important species, protected and safeguarded by local and community plans and regulations, on the other hand it has been interpreted as instability, and has contributed to keeping the deltas far from processes of modernization, relegating them to the role of "marginal periphery", places where for a long time nothing of relevance to the growth of the complex European economic systems has happened.

Today, however, the particular geomorphological and environmental conditions of these areas make them particularly sensitive to worsening climactic changes: rising sea levels (Nicholls Cazenave 2010), together with coastal erosion and increasingly intense and frequent droughts are contributing to a significant increase in the intrusion of salt water from the sea and to the consequent destruction of important agricultural produce and fishery production, as well as to local populations' difficulty to provide themselves with drinking water.

In the face of such risks, the development of technologically advanced devices designed to counteract them is no longer considered the most appropriate remedy: the fragility of these regions seems to require with increasing urgency an approach able to create a new balance between settlements, economic activities, the environment and hydraulic operations. (Collignon 2008)

Indeed, more and more an attempt is being made to mitigate the rigid control that for centuries the hydraulic infrastructure has wielded over these territories, trying to subordinate the ecological and environmental functioning to the agendas of the economy and of technology.

Faced with these issues, there is an attempt in the deltas to build a fertile meeting ground and place for mediation between environmental dynamics and anthropogenic processes, transforming the fragilities into opportunities through which to reinvent not only the delta landscapes and their hydraulic infrastructures, but also to outline new paths of more sustainable development. So on the one hand some are trying to renew the joining spaces between land and water, the great variety and articulation of wetlands that once dominated these territories, by transforming their status from areas deemed unproductive into new ecological-environmental-hydraulic infrastructures able to ensure the deltas' gradual adjustment to the threats posed by climate change; on the other hand there is experimentation with forms of sustainable farming yielding quality products, such as rice and vegetables in particular, or aquaculture and shellfish farming, and the integration of these forms of production with port and tourism activities attentive to the environment: this seems capable of ensuring a strong economy able to harness the considerable environmental resources. It is important to stress that the awareness that the excessive technical control and the hardening of the deltaic areas can be considered a cause – possibly the main cause – of their malfunction and inability to cope with climate change, is necessitating a substantial paradigm shift: the return to an adaptive approach that makes deep and detailed knowledge of the overall dynamics of the territory the key to correctly interpret and intervene in the deltas, and in which uncertainty is retained one of the important factors to be considered in devising flexible intervention measures and developing future scenarios (Ministry of Infrastructures and Environment 2011). Since this approach makes reintegrating the natural dynamics a strong point of the new planning for the deltas, water is transformed from an enemy to be fought into a potential ally through which to rearticulate the overall organization of these territories. As a result, it is the logics of water, whether related to dimensions of settlement, the economy, the environment or to technical aspects, that are playing an increasingly important role in the measures developed by the governments of various European deltas. However, if these measures are not integrated with each other, it could lead to intractable conflicts, for instance: attempting to ensure sufficient quality and quantity of drinking water for people living in the delta, but at the same time ensuring that the river flows are sufficient to combat the intrusion of salt water from the sea, while trying at the same time to ensure a sufficient provision of water to irrigate the farmlands in times of drought.

In particular, some of the issues identified as priorities by the various European deltas require deepening the knowledge of the inner workings of the territories and testing innovative solutions. A short list of issues may help to clarify their relevance also for other areas characterized by various or lesser fragilities.

1. *Freshwater supplies.* In addition to the cognizance that the availability of fresh water is one of the urgent needs to be placed on the agendas of local governments, it has also been understood how in these areas the deficiency related to the occurrence of exceptional droughts is becoming more and more a matter of routine: phenomena that rarely occur today will, in the future, increase their frequency, pushing us to experiment with socially sustainable and cost-effective solutions for the supply of freshwater. The development of reservoirs for seasonal rainwater or river water that at the same time can increase the presence of wetlands and ecosystems through which to purify the water and help animal species settle there is one of the main fields of experimentation.

2. *Multifunctional flood defense systems.* One important development concerns safety devices, particularly dams and embankments. These are increasingly being thought of as multi-functional elements, able to carry out a role of defense, but also one of ecological-environmental enhancement, besides being able to increase the allocation of spaces for leisure, in rural-urban situations where these elements are not always present.

3. *Natural flood risk management measures*. The management of flooding is increasingly expected to be done through actions involving the complex ecosystems of wetlands.

In this respect, the traditional strengthening of dams and embankments (elevation and extension) is being abandoned, in favour of a plurality of interventions including, for instance, moving the banks of canals and rivers, and the relative lowering of the alluvial areas to give more space to the flow of the river, the re-flooding of reclaimed land and the creation of canals for flooding.

4. *Hydro dynamism*. The reactivation of the full relationship, flooding, sedimentation becomes the precondition for lagoon areas to function once more as "natural sponges", absorbing flood waves and ensuring long-term water safety to the surrounding areas. In this way, in addition to defining a new structure of hydraulic protection, these spaces increase the accessibility and environmental quality of the territory by providing new, multifunctional wetlands, available for agriculture, aquaculture, tourism and leisure. This short list of topics describes only some of the main areas of experimentation of a very different approach, where integration and multi-functionality make the difference, ushering in a season of plans, projects and policies that at the same time address environmental fragilities, economic efficiency and spatial quality.

Since it is apparent that no single discipline can offer the solutions to such complex and articulated problems, many have called for an effort aimed at supporting the combining of knowledge, know-how and practices, to support that interaction of roles and disciplines that now seems able to improve the resilience of these territories.

This circulation of ideas and knowledge, so strongly requested, attributes a significant role to the knowledge produced by local societies, in defining intervention strategies and measures. In this sense, it is very important to emphasize the fundamental change that has occurred in the perception of risk on the part of the people who inhabit the delta. In fact, events that until the recent past were judged as an unavoidable fate by poor and backward populations, are now perceived as negligence, technical incompetence and lack of policy foresight by populations that utilize the search for paths of sustainable development and innovative security conditions as the tools to transform the fragilities into opportunities, outlining actions and intervention strategies able to put the liveability of these areas at the heart of the matter.

Taken together, these reflections reveal the great need for knowledge that these territories express, the important demand for these territories to be studied in depth, to be explored through projects and scenarios (Vellinga et al 2009), and at the same time to be contrasted and compared, in order to fill important gaps in knowledge on the one hand, and produce innovative methods of intervention on the other.

This book aims to contribute at least partially to filling this gap, trying to advance the knowledge and the know-how concerning territories as fragile as they are rich in opportunity.

References

- Collignon, R. (2008), "Room for the River", in Cecile va der Heijden, Michiel Veldkamp (ed), *Designing with water*, Janssen/PersRotatiedruk, Gennep
- www. delta-alliance.nl/deltas
- Bucx, T., Marchand, M., Makaske, A., van de Guchte, C., (2010), "Comparative assessment of the vulnerabilità and resilience of 10 deltas" – synthesis report, in *Delta Alliance report* 1, Delta Alliance International, Delft-Wageningen, The Netherlands
- Ministry of Infrastructure and Environment (2011), *Delta Programme 2012. Working on the Delta*, The Netherlands
- Nicholls, R.J., Cazenave, A., (2010), "Sea-level rise and its impact on coastal zones", *Science* 328, pp. 1517-1520
- Vellinga, P., et al (2009), *Exploring high-end climate change scenarios for flood protection of the Netherlands*. KNMI scientific report WR 2009-05, Royal Netherlands Meteorological Institute (KNMI), De Bilt

Part One

European Delta Landscape

Biography and comparison as instruments of knowledge

The difficulty of understanding a territory has often led researchers to experience it through tools and research practices based on direct observation. It is not uncommon, in fact, for the need to be felt in Western culture, in the face of major changes and transformations in the surrounding world, to exit the libraries in order to experience forms of knowledge that lie elsewhere, not in books (Blumemberg, 1989).

Among these, experiencing places directly, measuring them, describing them and comparing their essential characteristics are perhaps the main ways of appropriating this knowledge, by drawing heavily on the languages and analytic categories that arise from physical experience. (Zumthor, 1993)

Dense/sparse, far/near, big/small, empty/full are just some of dichotomous pairs derived from the experience of a physical body in space that enable us to recount and unveil a territory's characteristics. Drawing maps of the area, using these and other categories, can reveal surprising connections, and also dispel false images or common rhetoric: in this sense it can produce new knowledge for formulating interpretative hypotheses and for defining intervention strategies.

Since obtaining specific and precise knowledge of each area of the European deltas seemed as urgent as producing maps to explore and recognize similarities and differences between these territories, two complementary forms of description have been employed. A "thin" and abstract description through which to highlight everything these areas have in common (Walzer, 1994), and a "thick" description (Geertz, 1973) capable of illustrating the specificity and articulation of the history and culture of each delta.

Thus, on the one hand, comparing territories through mapmaking, confronting them and examining them side by side has proved useful to consider their differences or similarities, and to establish a mutual relationship between territories that allows their evaluation; a tool to place specific situations within wider and more general horizons. On the other hand, it was deemed important to write the biography of each delta territory in order to recount and narratively restore its specific processes of formation and transformation rather than its ultimate and seemingly stable configuration; a tool capable of acquiring knowledge that goes beyond the presumed processes of territorial standardisation, and that through the reconstruction of specific local descriptive-interpretive frameworks was judged capable of showing how similar territorial situations may be the result of very different development paths.

Despite the specificity of each of the cognitive instruments used, comparison and biography have been employed in close conjunction; thus, the few general interpretations of the deltas we thought it possible to formulate have been developed through a detailed exploration of each individual location.

These two in-depth investigations and comparison activities, taken jointly, were considered an important first step towards the realization of an Atlas of the European deltas. An Atlas understood as a composite device with different aims and purposes: to reveal little known realities which have remained on the margins of the debate concerning the transformations of the European territory; to bring to the fore the occurrence of some of the problems and weaknesses of these areas and the plurality of strategies used to deal with them, often in an uncoordinated manner; to support decision making by constructing possible frames of reference.

The Danube in Romania, the Ebro and the Guadalquivir in Spain, the Nemunas in Lithuania, the Po in Italy, the Rhine in the Netherlands, the Rhône in France are the rivers whose outlet to the sea we have studied. This group of deltas was chosen based on some considerations related primarily to the moderate-to-low density of the populations inhabiting these areas (with the exception of the Rhine delta, which though densely populated never reaches the levels of the Asian deltas), the characteristics of the seas on which they are located, the blend of agricultural and tourism functions found there and, finally, the availability of information and the presence of institutions interested in collaborating in the study.

At the outset we felt it important to initiate the

comparison of these territories by moving precisely from the uneven levels of knowledge, studies and deepened research already accumulated for each delta. Some of these deltas are better-known, and play a clearer role in the collective imagination: about them we collectively assume to have a clearer picture; others are much more marginally represented in studies and research. For this reason, the confrontation and comparison also served to develop units of measurement through which to place the characteristics of each territory against the backdrop of the others, initiating a process of understanding what we don't know through what we do.

This type of work saw the direct involvement of universities and also of technicians working for the government institutions of the delta territories under comparison, in order to highlight themes and issues, and ongoing projects. The materials and information collected were processed with the aim of creating maps of synthesis, diagrams and descriptive charts of each delta analyzed, but also to focus on that set of issues capable of outlining the common characteristics distinguishing these territories as a group.

Despite the deltaic regions in Europe generally being assailed by some significant weaknesses, including the more serious issues of the salinisation of the water, coastal erosion and rising sea-level, the comparison made some phenomena recognizable that strongly signal the singularities of each region, such as, for example, the gradual abandonment of the land by the local population in the Po Delta, the strong increase of the pressures of tourism in the deltas of the Rhone and the Guadalquivir, the transformation and strengthening of agriculture in the Ebro delta, the high level of urbanization of the Rhine delta, or the isolation and the predominance of untouched nature along the Danube delta. Similarity and difference are the keywords that guided the creation of some series of maps used to describe these European deltas, tracing their main characteristics and contributing in this way to identifying strategies and effective measures to address critical issues. Stated differently, the effort was to show how a plurality of themes, problems and phenomena have traced different geographies across the European region, each of which has required and continues to require specific types of intervention. Benefiting and learning from the experience of others and simultaneously measuring the difference between similar areas on the basis of their ability to imagine their own future have also been instrumental to this end.

The survey selected themes capable of highlighting important ecological issues, mainly linked to hydraulic risks and climate change, but it also chose specific economic and social problems affecting the delta territories, which bring out a set of common problems. The main themes concern: the management of the hydrographic network, the loss of local economic potential due to the risk of flooding, the agricultural crisis, especially for niche products such as rice, due to salinisation and the increased frequency of droughts, the disappearance of natural habitats and loss of biodiversity. Risks which are in turn linked to the increase in average global temperatures.

More generally, the thematic field within which the instruments of biography and comparison were used can be reduced to six main issues.

1. Limits

A first set of questions covered the different size and geography that each delta traces upon the territory. What are the scales of reference and limitations (geographic, hydrographic, settlement, economic, environmental, social, administrative, etc.) of the selected delta territories? How do the geographic limits change depending on the viewpoints chosen and the reporting systems considered? To which territorial context do the strategies, plans and projects in place refer?

2. Infrastructure

This issue involved asking questions about the infrastructure systems present in the territory: roads, railways, port areas, waterways, and so on. Which networks have taken shape? With what systems is there a relationship of mutual influence (environmental systems, natural systems, tourist areas, productive systems, trade and economic systems, etc.)? How are the flows articulated and how are resources placed in relation to one another? What are the main infrastructures (roads, railways, hydraulic, etc.) and through which areas do they pass?

3. Landscapes

A third issue regarded the environmental and ecological systems present in the territory: wetlands, lagoons, dunes, floodplains, marshes, fishing ponds, irrigation systems, catch basins of fresh water, reclamation areas, plant systems, forests, reeds, etc. To what recurring landscapes has the blend of these different materials given rise in the deltaic areas considered? Are they stable landscapes or situations in continuous evolution and transformation?

4. Development and protection

Research into this issue involved examining the planning systems and projects that govern these territories, which entities and institutions they act upon and over which portions of land they have jurisdiction, etc.

What tools and control policies have been implemented in recent years?

Through which plans and projects (protected and restricted areas, infrastructural and urbanization solutions, energy plans, etc.) is an attempt being made to guide the transformation of the territories in analysis?

What weaknesses and risks (saltwater intrusion, coastal erosion, subsidence, pollution, urbanization, etc.) affect the selected territories? What actions are being undertaken to remedy to these?

5. Habitats and cultures

A further issue relates to settlement systems, residential systems (historical centres, reclaimed houses, recent subdivisions, urban roads, resorts, etc.), manufacturing systems, services and equipment, etc.

What new forms of living (residential, work-related, related to the provision of services or of hospitality, etc.) are developing in relation to new lifestyles and new populations?

6. Tourism and leisure

A final theme of investigation focused on tourism infrastructure and the materials of which it is composed, organized beaches, rest points and observatories, hotels and resorts, tourist centres, tourist routes, systems of bike paths, oases and visitor centres, etc.

What dynamics related to tourism and leisure are affecting the territories under study?

Comparison and biography have been employed throughout this set of questions, in an attempt to organize them, as has been done in other studies on deltaic areas (Bucx, 2010), in three large realities: the Occupation layer (zoning of land and water use), the Network layer (infrastructures, dykes and dams, etc.) and the Base layer (natural resources, natural coastal defence structures such as beach ridges and dunes, etc.), each with different but interrelated temporal dynamics and public-private involvement. The need to improve the welfare and living conditions of the people who inhabit the delta areas which more than other areas are subject to risks and criticality; the urgency to collect data and information to help improve the quality of the projects and plans proposed for these territories; the possibility of extending the knowledge acquired and disseminating it among researchers and entities that govern these territories.

The effort to start the creation of an Atlas of European Deltas is a first tentative answer to this set of needs.

References

- Blumemberg, H., (1989) *La leggibilità del mondo*, Il mulino Bologna (ed. or. 1981)
- Bucx, T., Marchand, M., Makaske, A., van de Guchte, C., (2010), "Comparative assessment of the vulnerabilità and resilience of 10 deltas", Synthesis report, in *Delta Alliance report* 1, Delta Alliance International, Delft-Wageningen, The Netherlands
- Geertz, C., (1973), *The Interpretation of Cultures*, Basic Books, N.Y.
- Zumthor, P., (1993), *La mesure du monde*, Éditions du Seuil, Paris

Geographies and Comparison

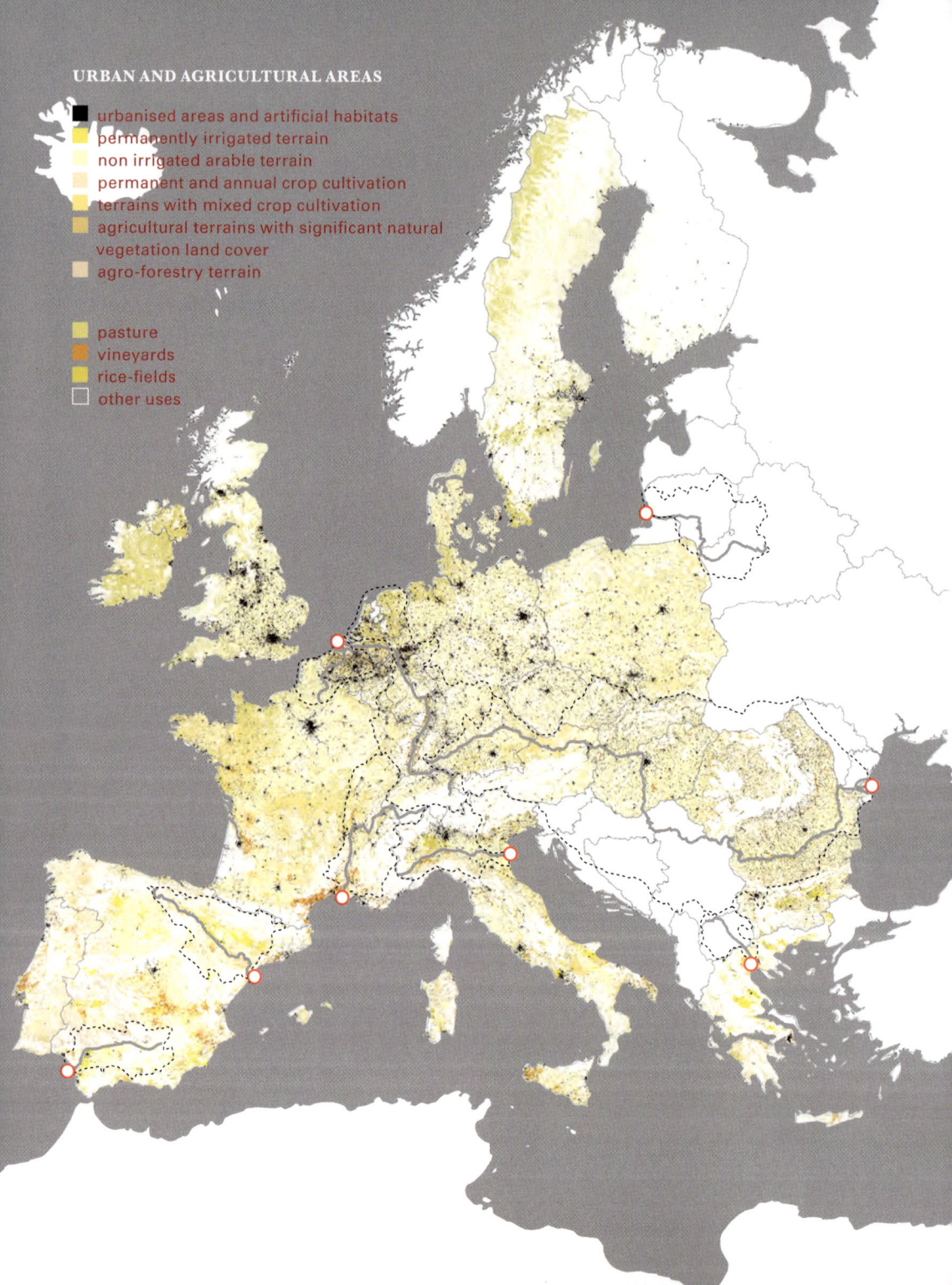

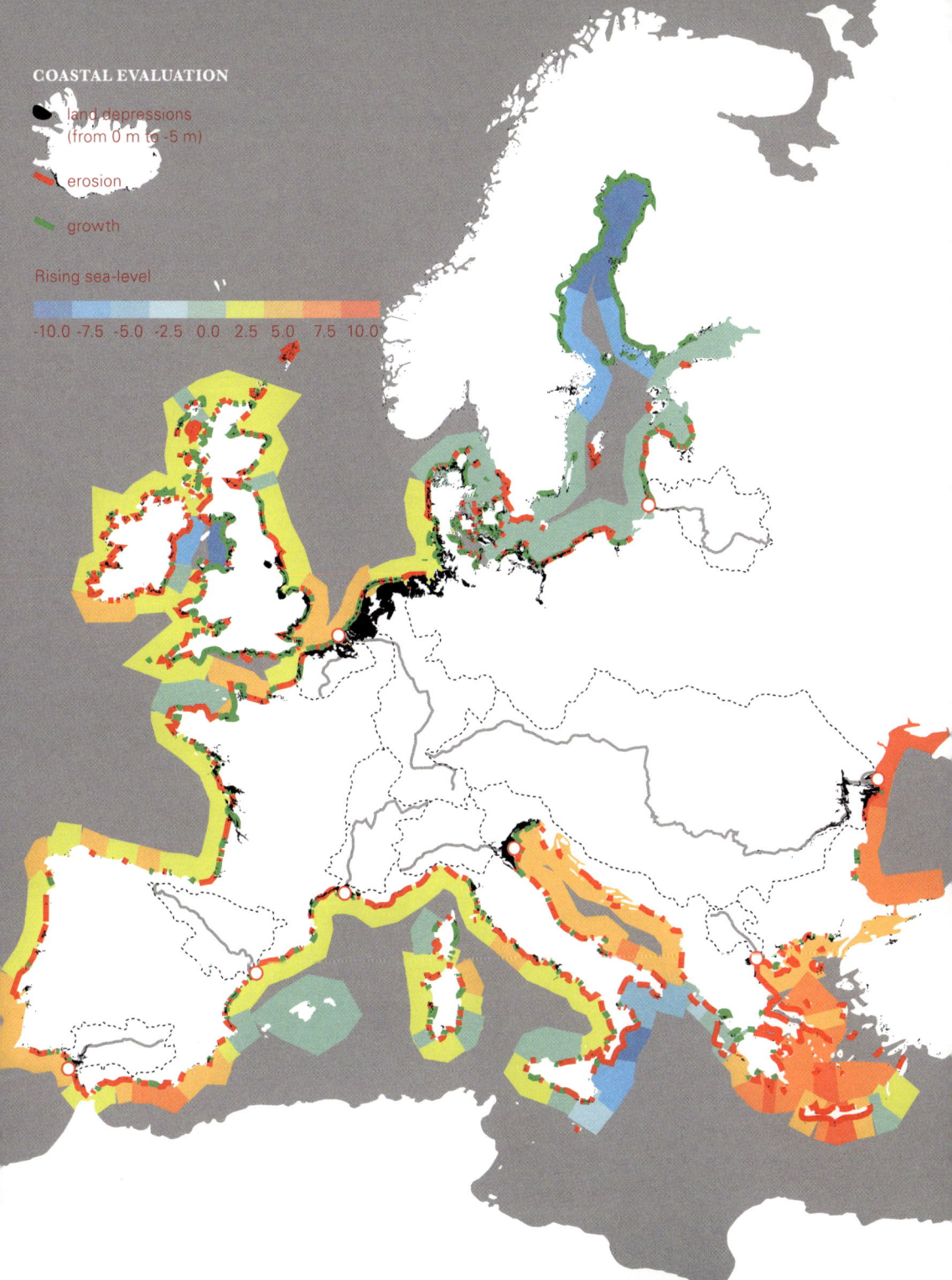

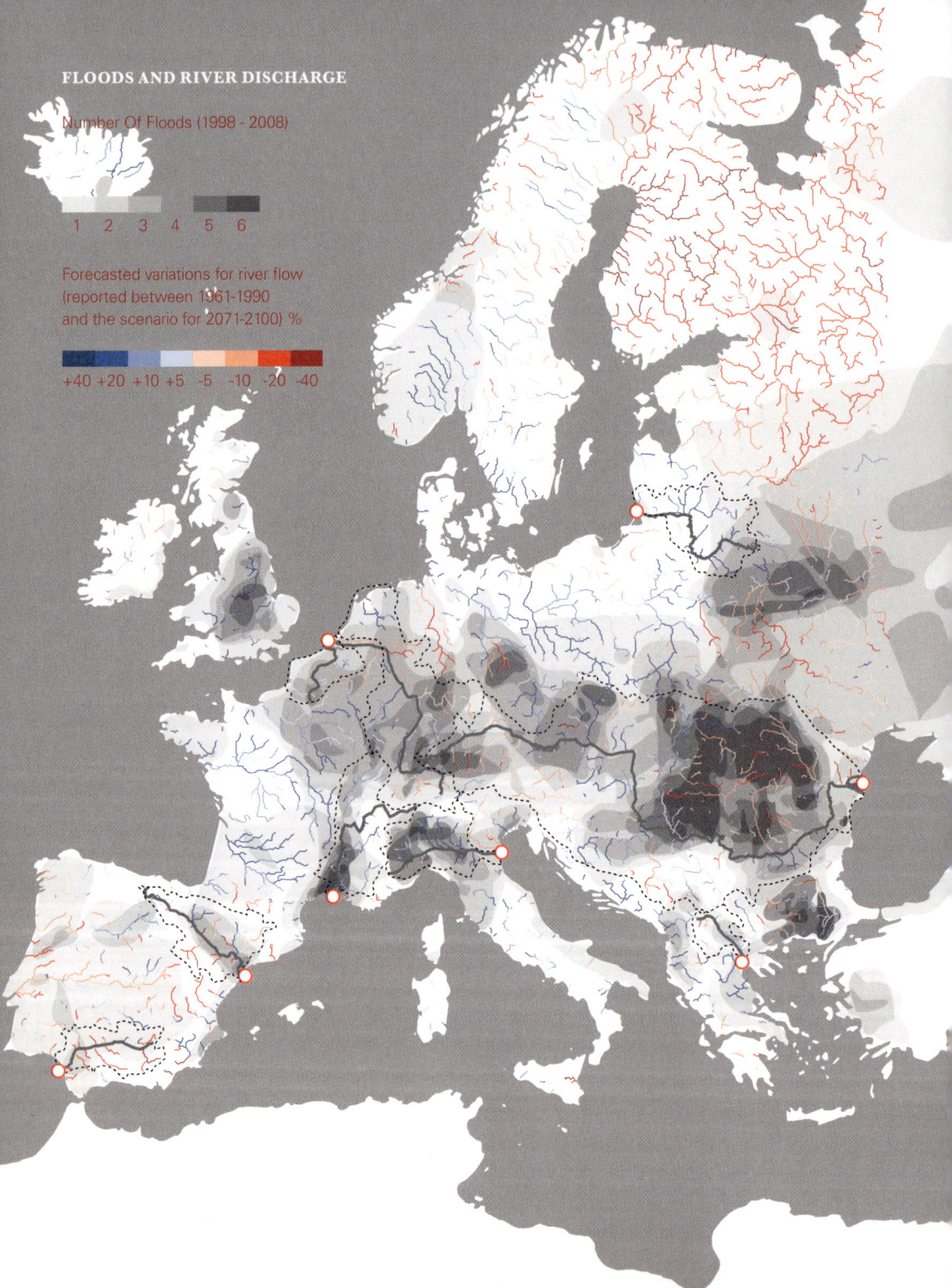

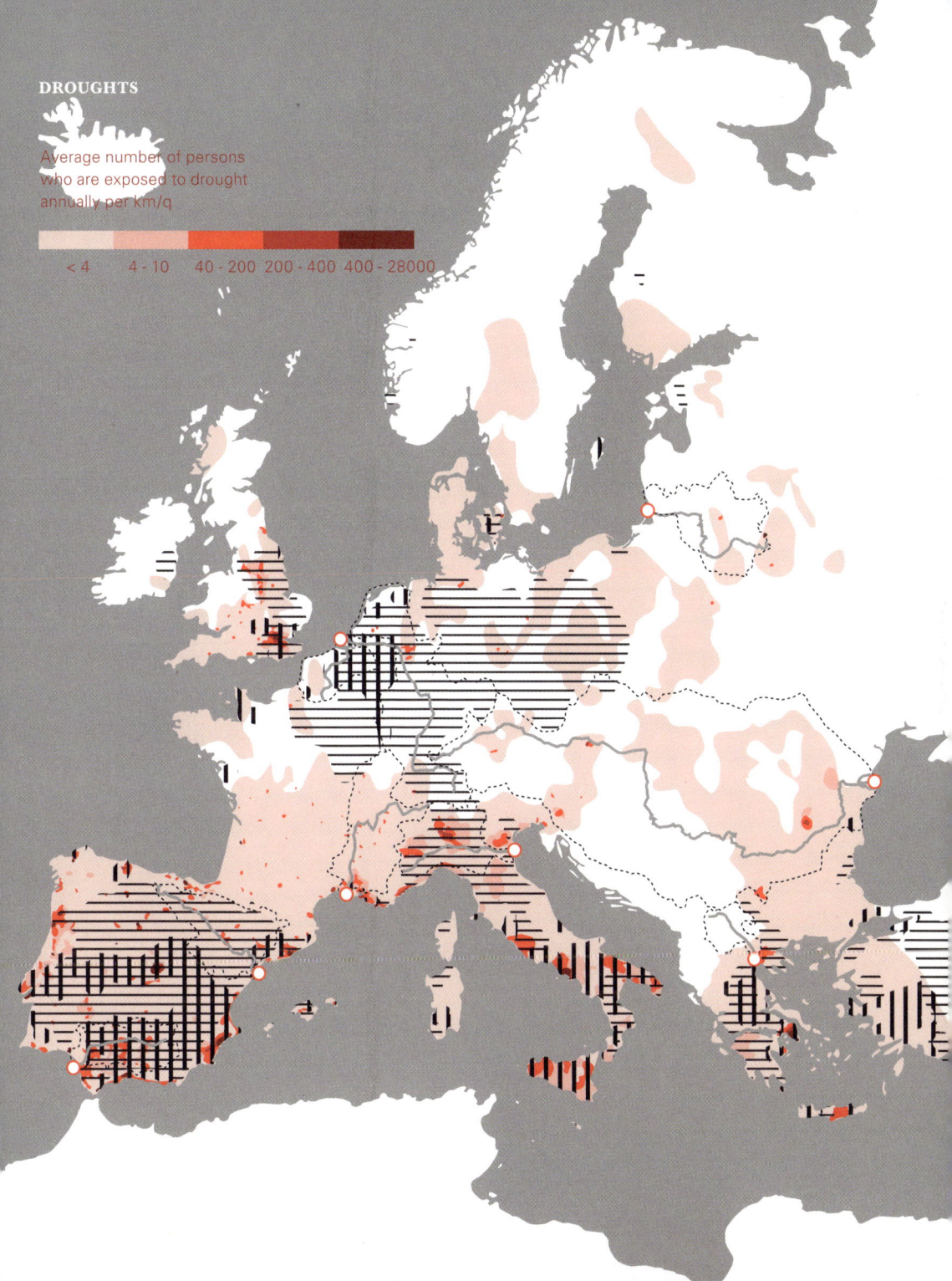

RHINE

300 km

NEMUNAS

18 km

EBRO

33 km

RHONE

40 km

PO

40 km

GUADALQUIVIR

58 km

DANUBE

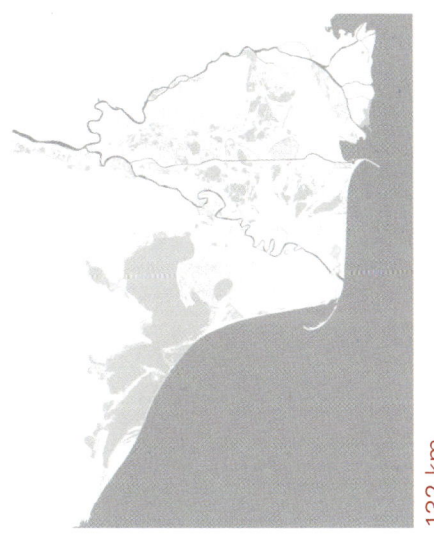
132 km

DELTA SCALE

NEMUNAS

EBRO

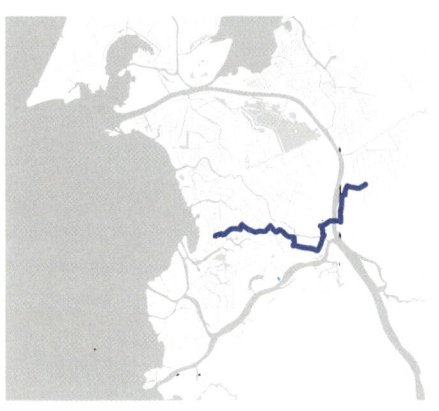

GUADALQUIVIR

PO

RHINE

DANUBE

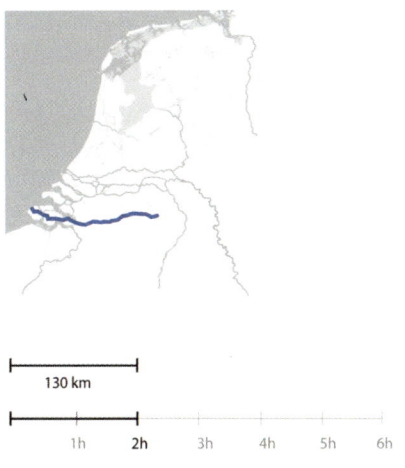

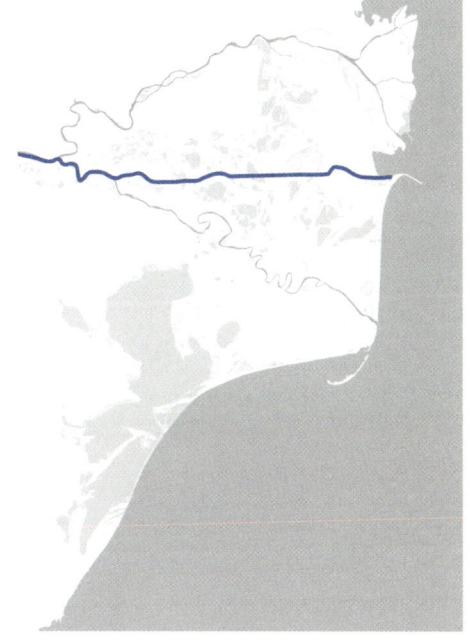

RHONE

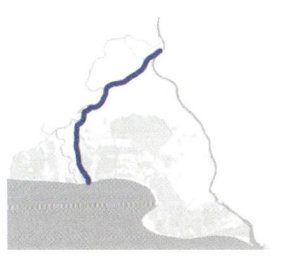

TIMESCALE

ATLAS - Part One | European Delta Landscape | Geographies and Comparison

NEMUNAS

area: 239,5 kmq
branches: Nemunas
population: 53.376 inhabitants
density: 31 inhabitants/kmq
mouth: Baltic Sea

NEMUNAS RIVER
lengths: 937 km
discharge average: 616 m3/s
basin: 98.000 km2
countries: 3

EBRO

area: 330,31 kmq
branches: Ebro
population: 55.000 inhabitants
density: 50,5 inhabitants/kmq
mouth: Mediterranean Sea

EBRO RIVER
lengths: 910 km
discharge average: 426 m3/s
basin: 80.093 km2
countries: 1

GUADALQUIVIR

area: 543 kmq
branches: Guadalquivir
population: 180.000 inhabitants
density: 25 inhabitants/kmq
mouth: Atlantic Ocean

GUADALQUIVIR RIVER
lengths: 657 km
discharge average: 164.3 m3/s
basin: 56.978 km2
countries: 1

PO

area: 787.29 kmq
branches: Po di Pila, Po delle Tolle, Po di Donzella, Po di Goro, Po di Maistra
population: 73.119 inhabitants
density: 93 inhabitants/kmq
mouth: Adriatic Sea

PO RIVER
lengths: 625 km
discharge average: 1540 m3/s
basin: 74.000 km2
countries: 2

RHINE

area: 41.543 kmq
branches in Netherland: Waal, Nederjin/
Let, Ijssel
population: 16.458.787 inhabitants
density: 397 inhabitants/kmq
mouth: North Sea

RHINE RIVER
Lengths: 1320 km
discharge average: 2200 m3/s
basin:170.002 km2
countries: 8

DANUBE

area: 5.165 kmq
branches: Sulina, Chilia, Sfantu Gheorghe
population: 123.878 inhabitants
density: 30 inhabitants/kmq
mouth: Black Sea

DANUBE RIVER
lengths: 2850 km
discharge average: 6500 m3/s
basin: 817.000 km2
countries: 10

RHONE

area: 930 kmq
branches: little Rhone, big Rhone
population: 110.000 inhabitants
density: 10 inhabitants/kmq
mouth: Mediterranean Sea

RHONE RIVER
lengths: 813 km
discharge average: 1.710 m3/s
basin: 98.000 km2
countries: 2

NEMUNAS

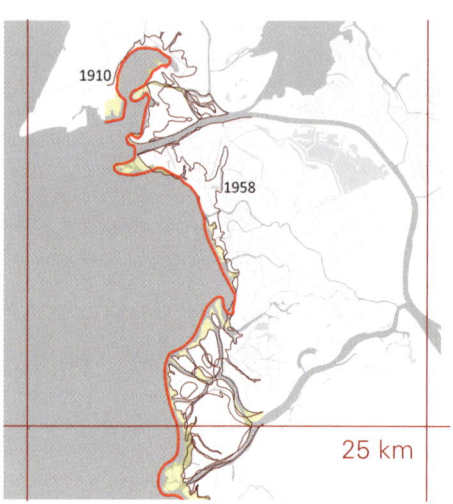

EBRO

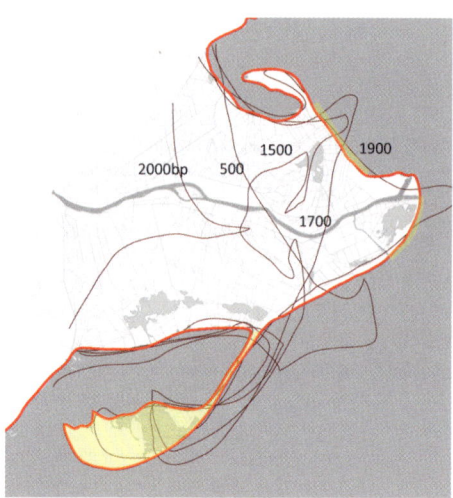

GUADALQUIVIR

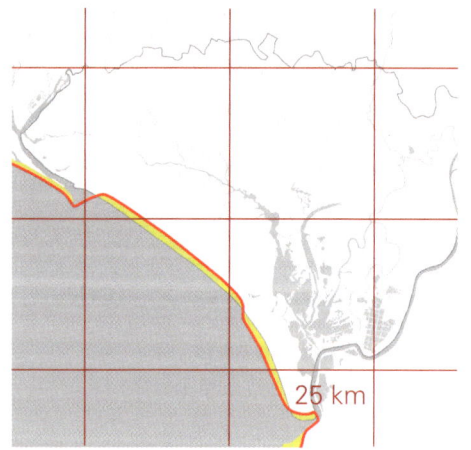

PO

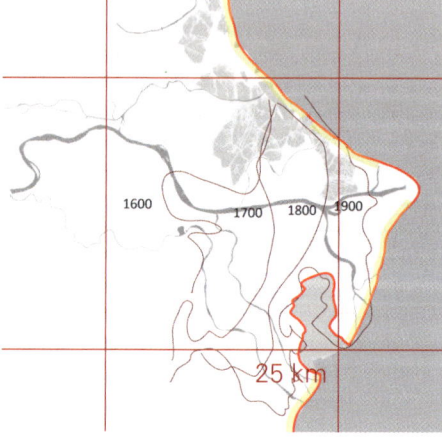

RHINE

DANUBE

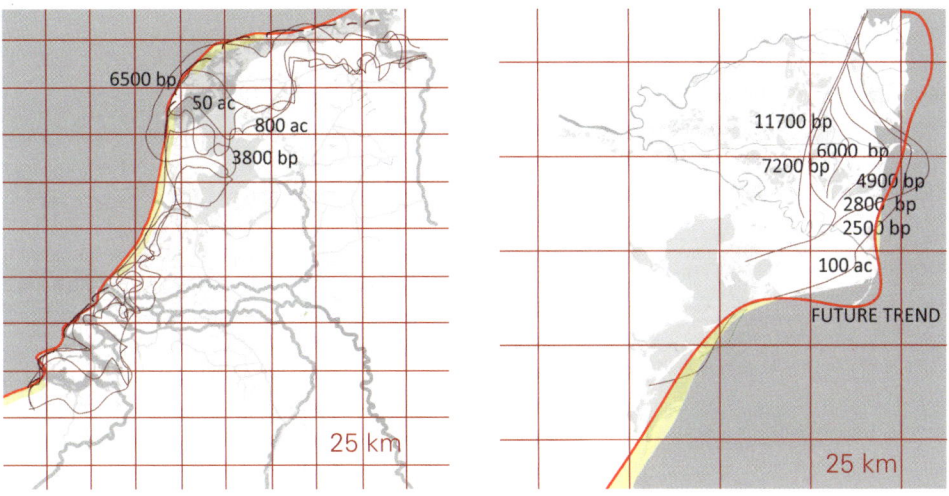

RHONE

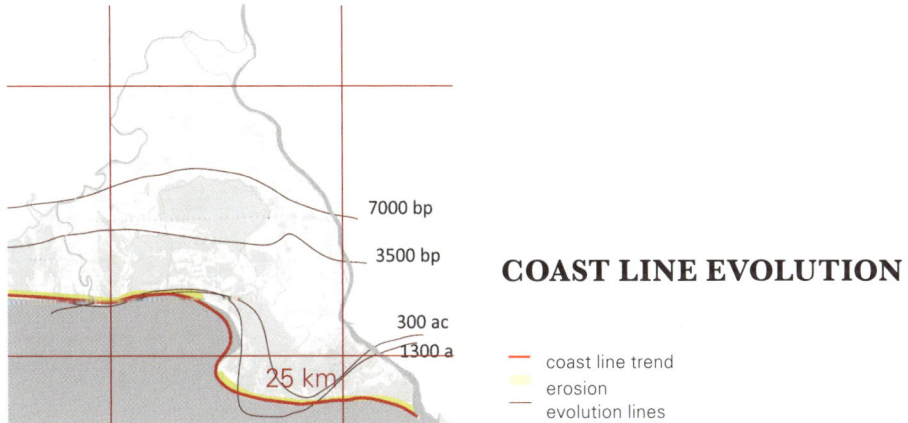

COAST LINE EVOLUTION

— coast line trend
▬ erosion
— evolution lines

NEMUNAS

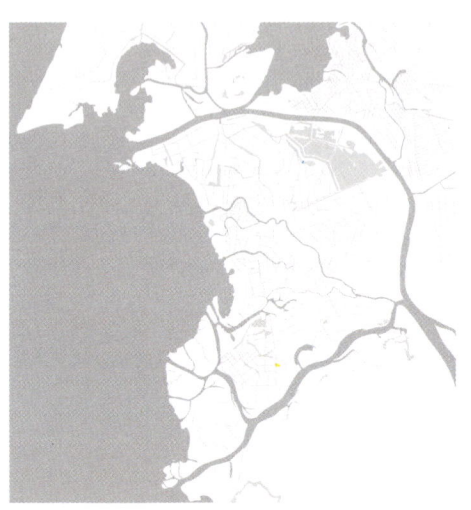

EBRO

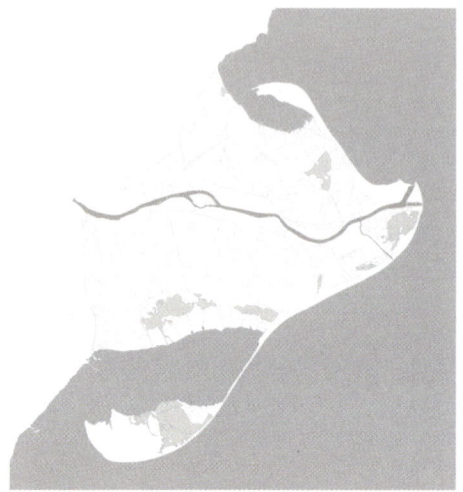

GUADALQUIVIR

PO

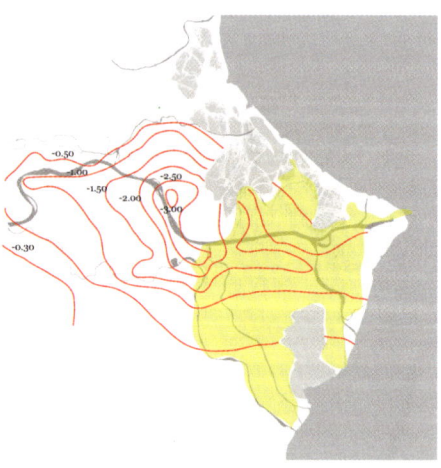

RHINE   **DANUBE**

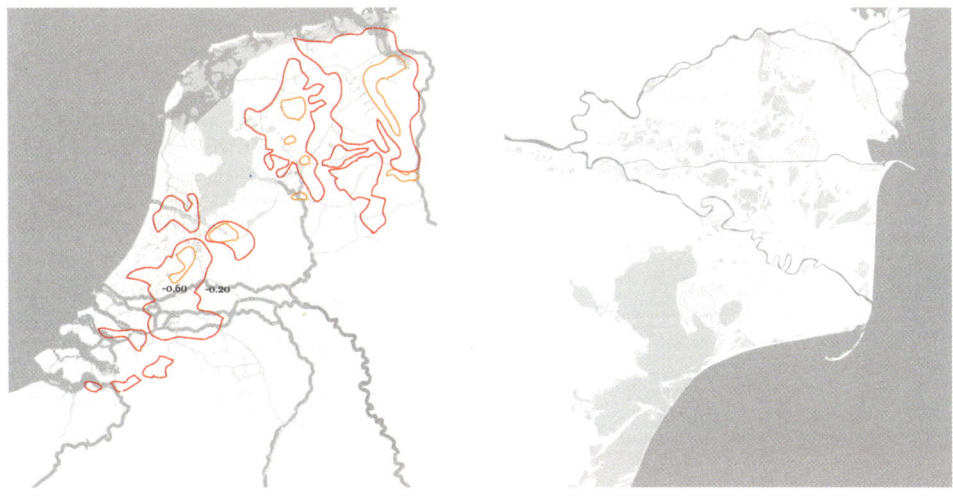

RHONE

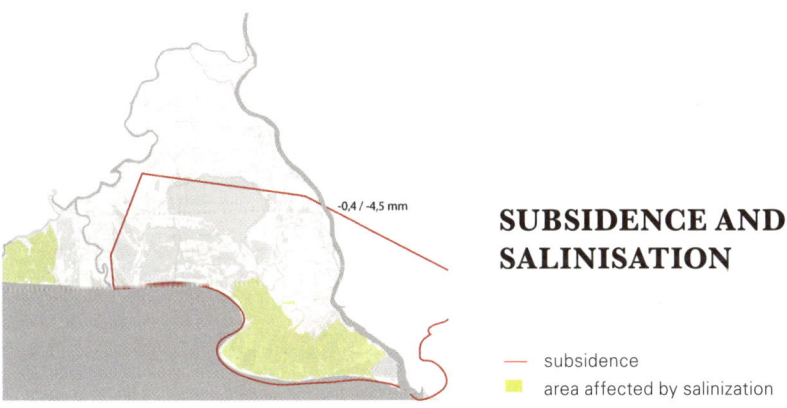

SUBSIDENCE AND SALINISATION

— subsidence
▪ area affected by salinization

ATLAS - Part One | European Delta Landscape | Geographies and Comparison

NEMUNAS

EBRO

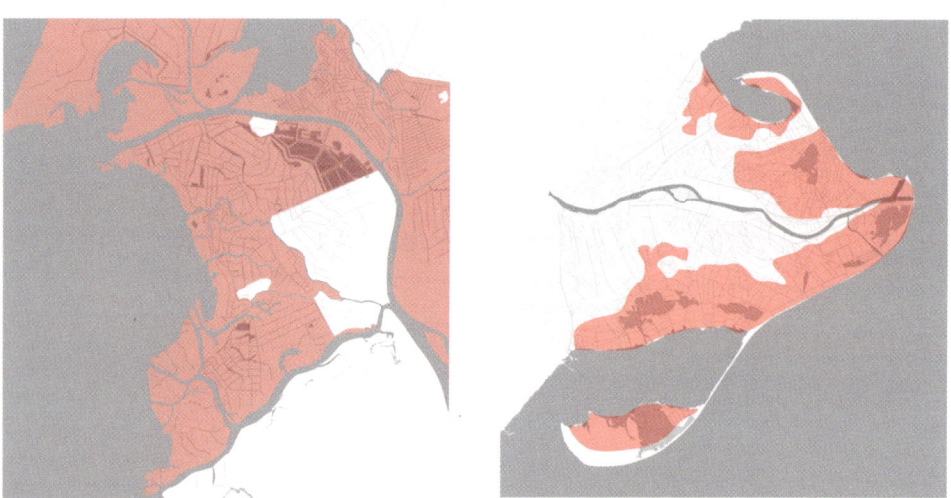

GUADALQUIVIR

PO

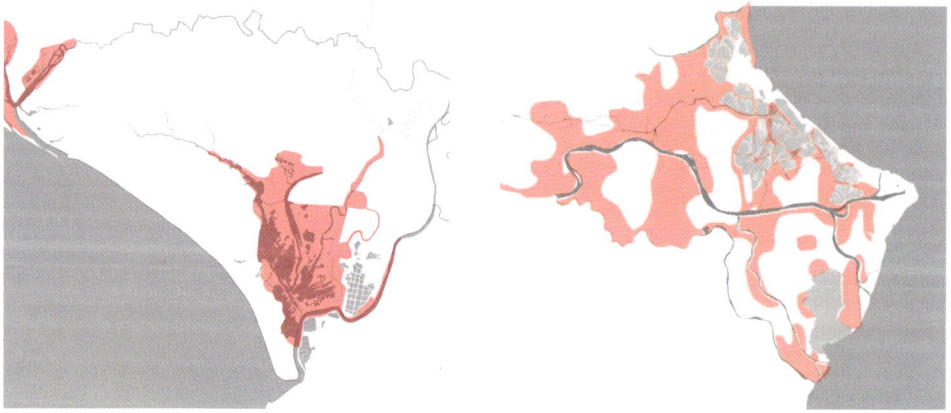

RHINE

DANUBE

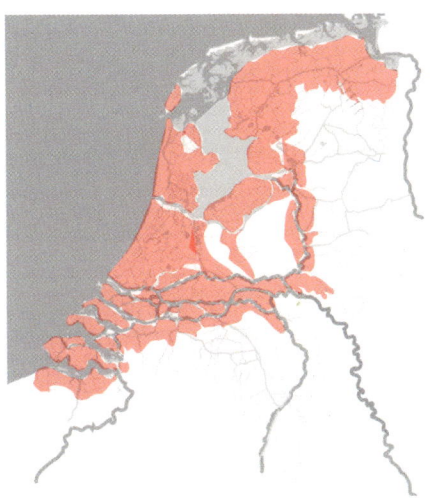
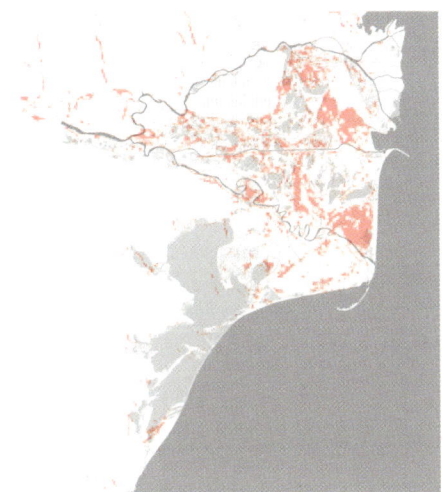

RHONE

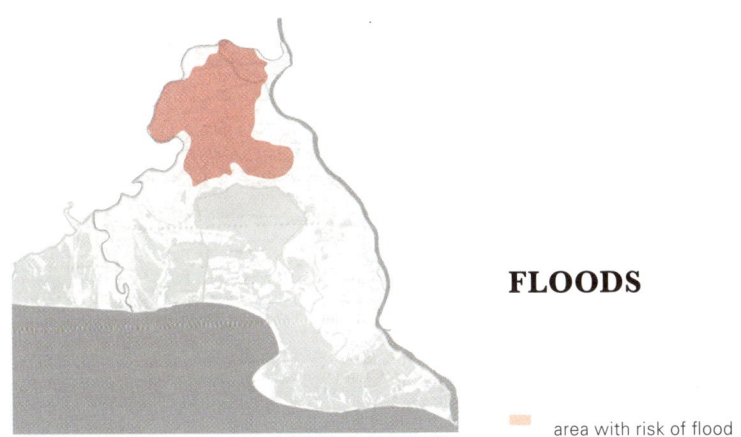

FLOODS

▬ area with risk of flood

ATLAS - Part One | European Delta Landscape | Geographies and Comparison

NEMUNAS **EBRO**

GUADALQUIVIR **PO**

RHINE **DANUBE**

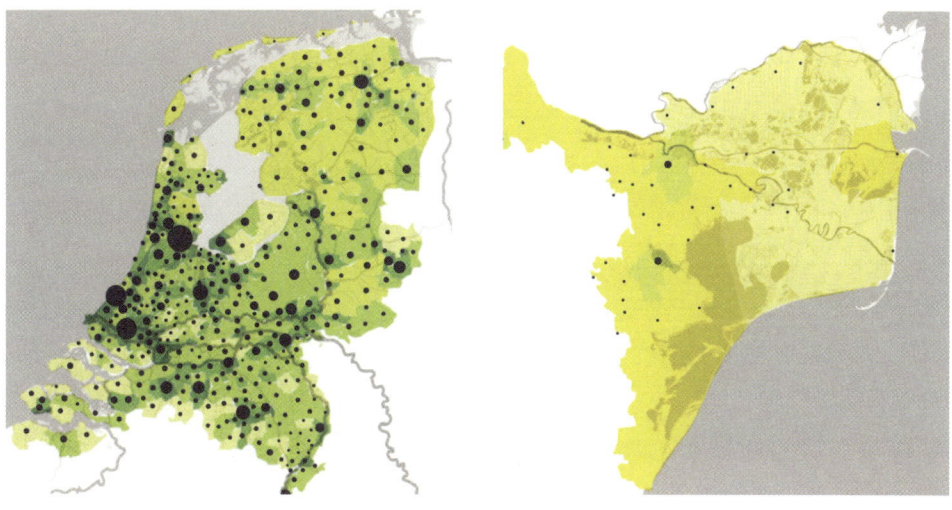

RHONE **SETTLEMENTS AND POPULATION**

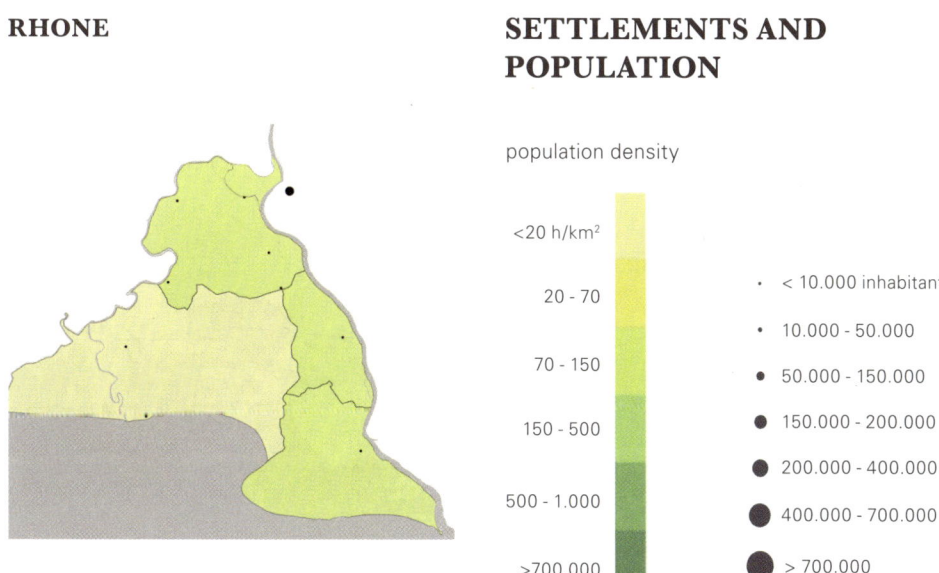

population density

- <20 h/km²
- 20 - 70
- 70 - 150
- 150 - 500
- 500 - 1.000
- >700.000

- < 10.000 inhabitants
- 10.000 - 50.000
- 50.000 - 150.000
- 150.000 - 200.000
- 200.000 - 400.000
- 400.000 - 700.000
- > 700.000

NEMUNAS

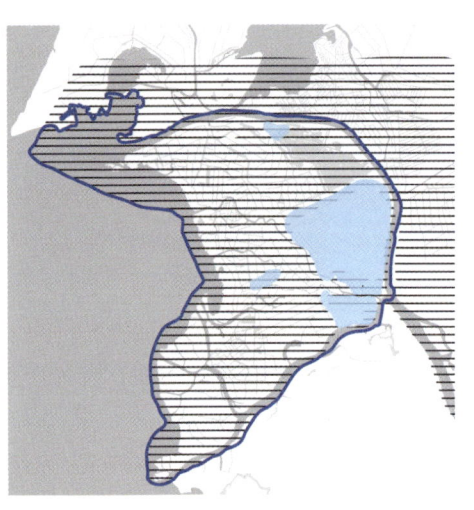

EBRO

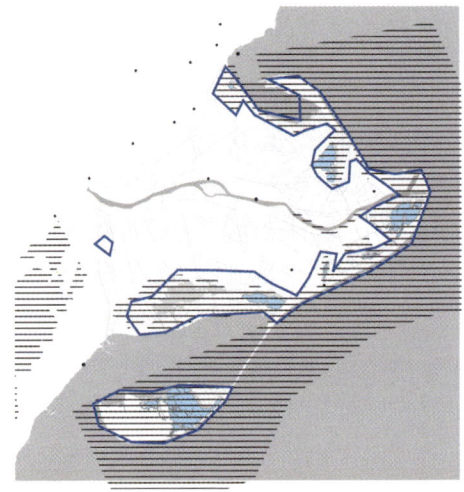

GUADALQUIVIR

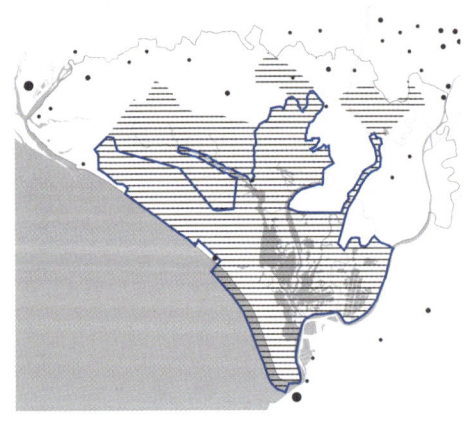

PO

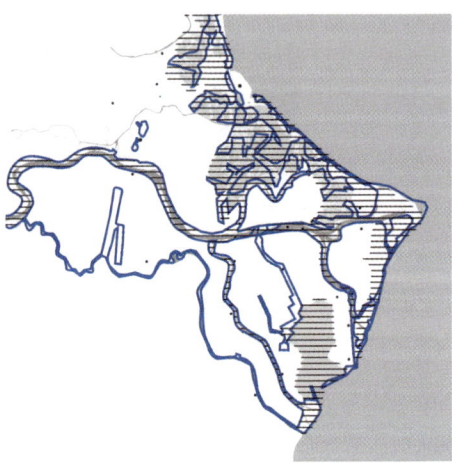

RHINE **DANUBE**

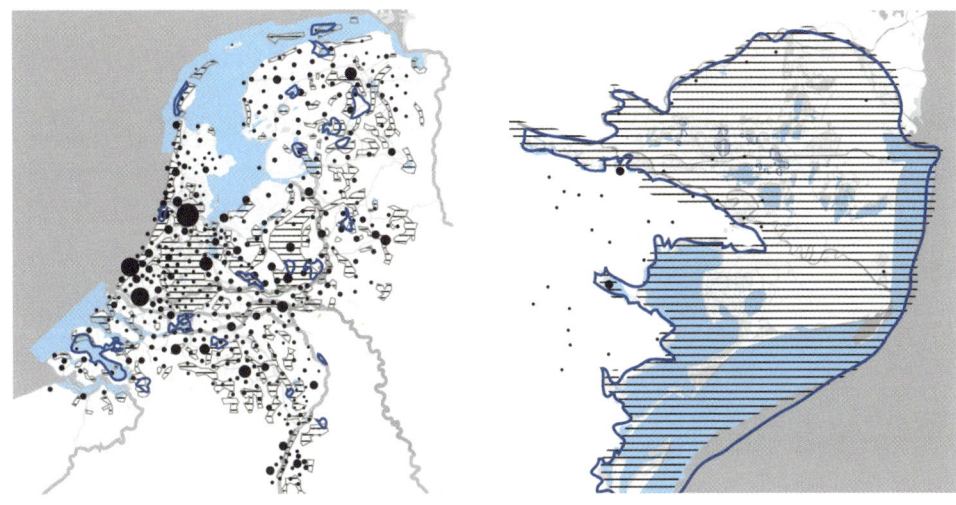

RHONE

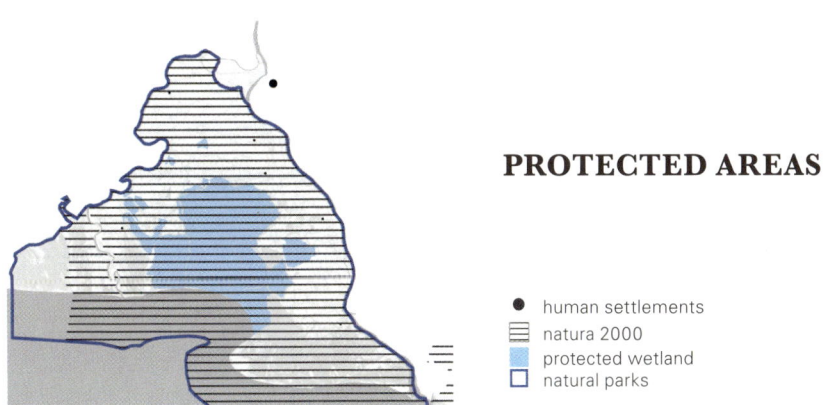

PROTECTED AREAS

● human settlements
☰ natura 2000
▨ protected wetland
☐ natural parks

ATLAS - Part One | European Delta Landscape | Geographies and Comparison

NEMUNAS

EBRO

GUADALQUIVIR

PO

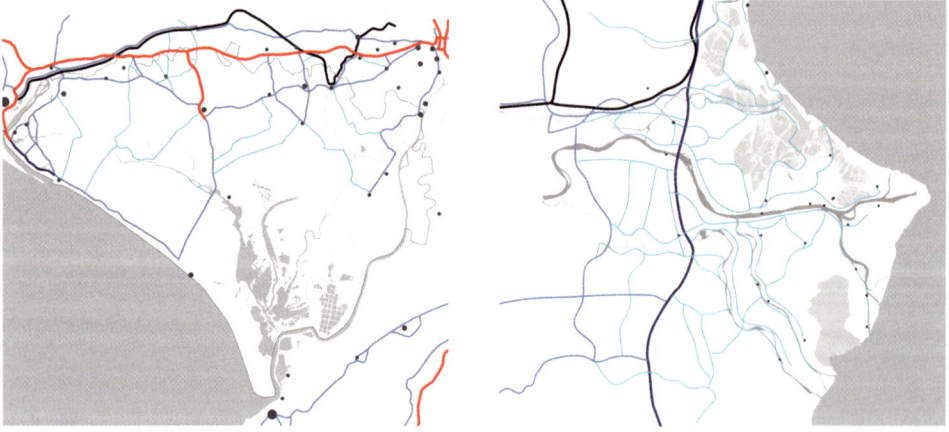

RHINE   **DANUBE**

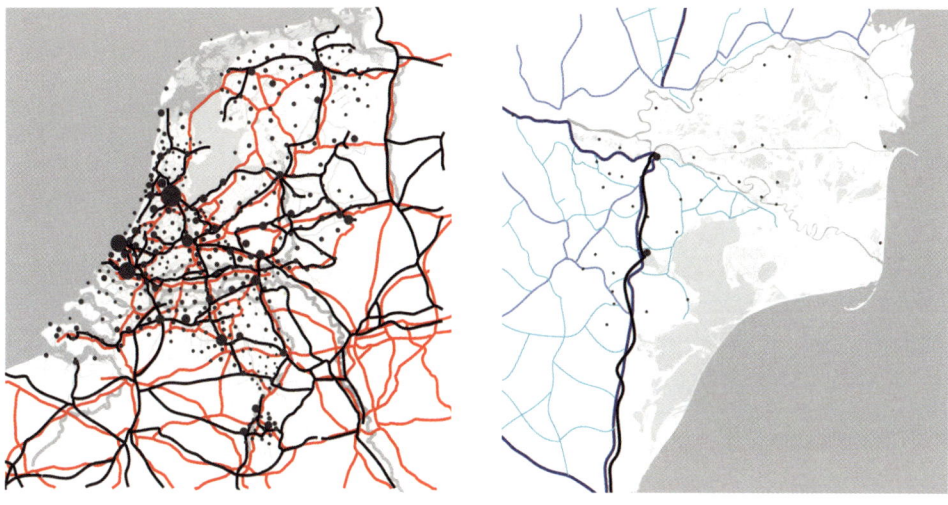

RHONE

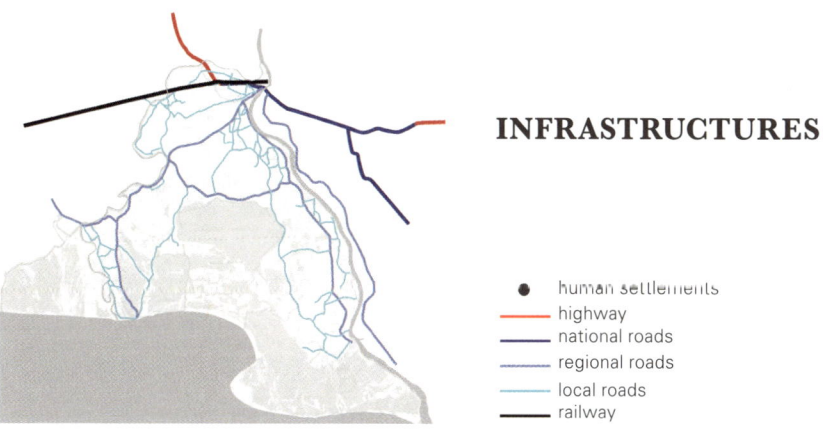

INFRASTRUCTURES

● human settlements
— highway
— national roads
— regional roads
— local roads
— railway

ATLAS - Part One | European Delta Landscape | Geographies and Comparison

NEMUNAS

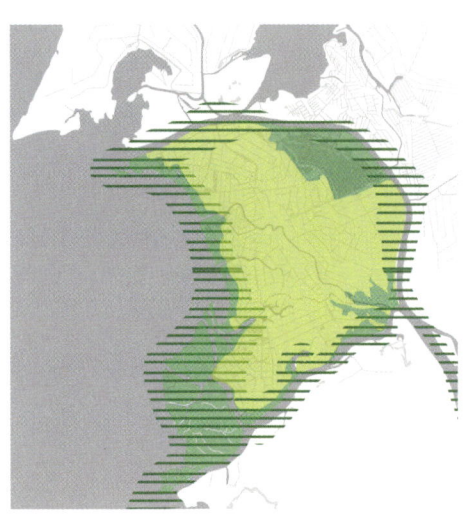

EBRO

GUADALQUIVIR

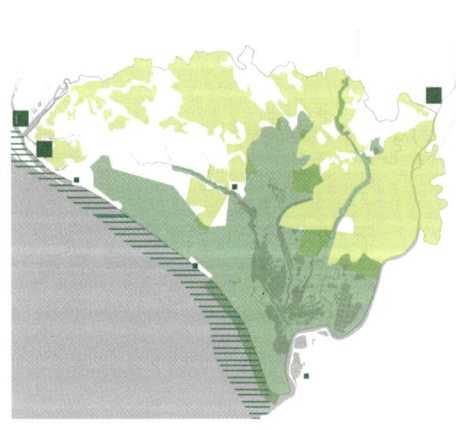

PO

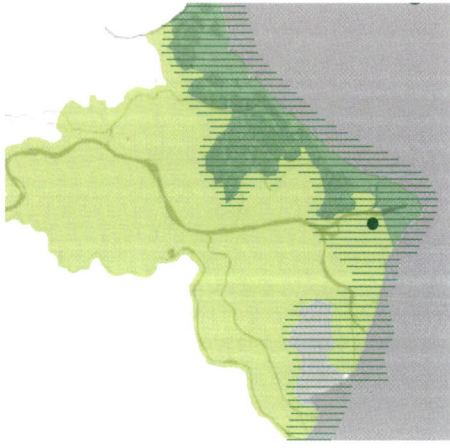

RHINE **DANUBE**

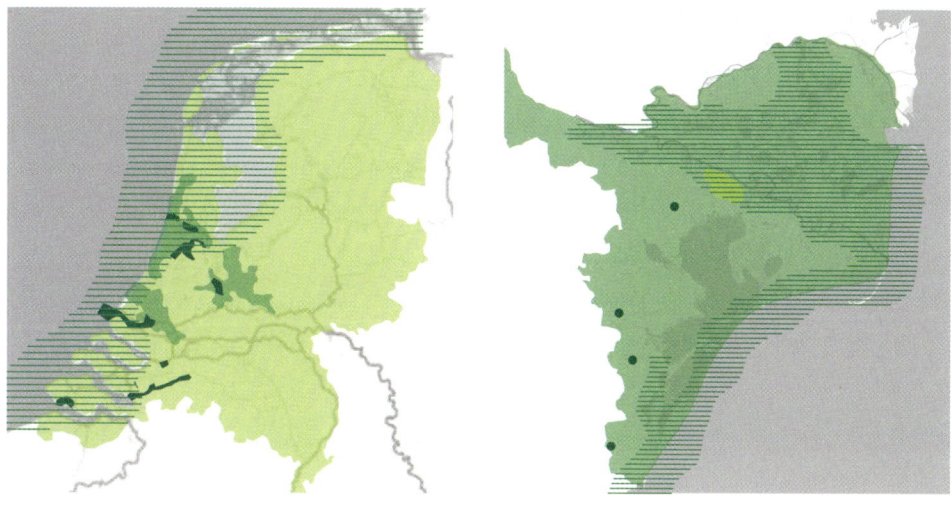

RHONE

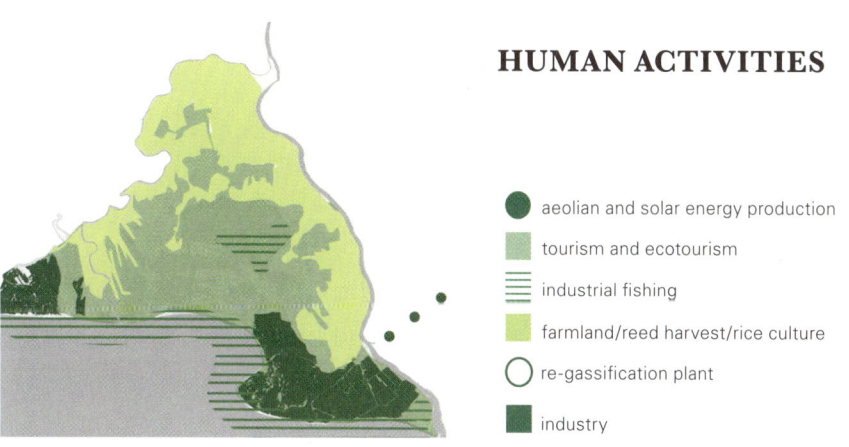

HUMAN ACTIVITIES

- ● aeolian and solar energy production
- tourism and ecotourism
- ≡ industrial fishing
- farmland/reed harvest/rice culture
- ○ re-gassification plant
- ■ industry

NEMUNAS

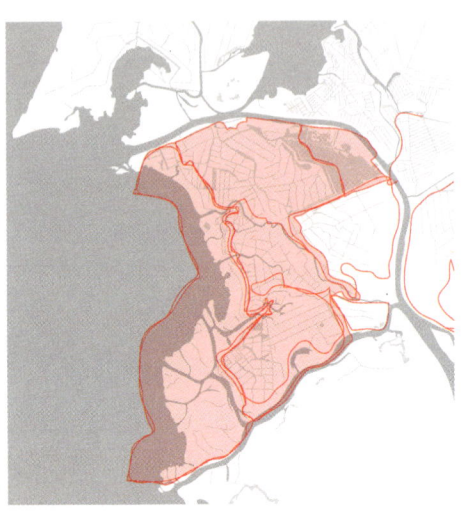

EBRO

GUADALQUIVIR

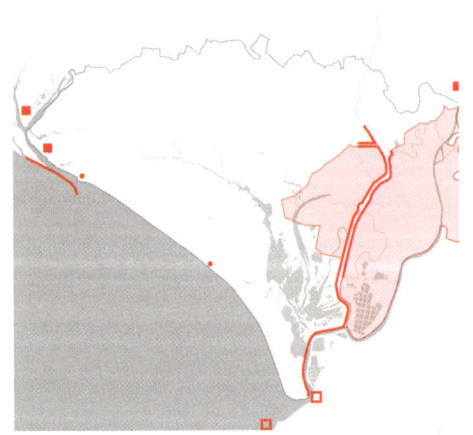

PO

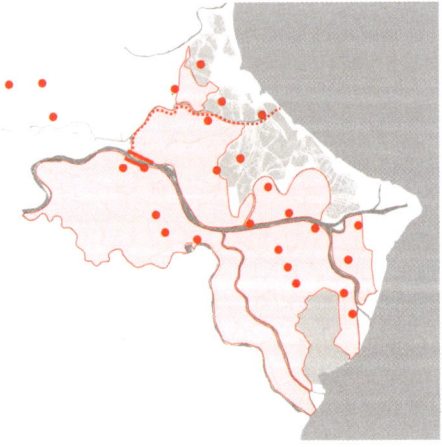

RHINE

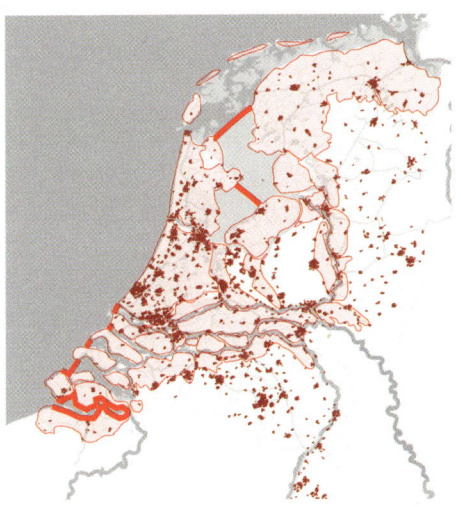

DANUBE

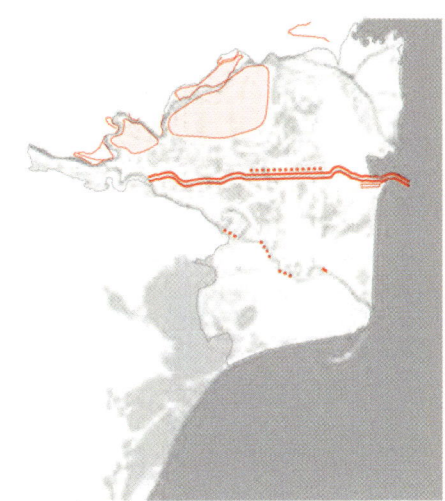

RHONE

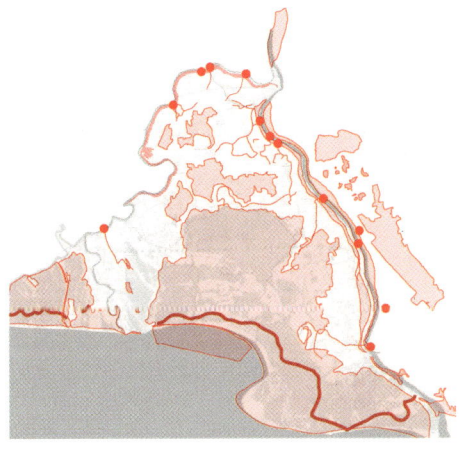

TRANSFORMATIONS

- ⬭ reclaimed lands/dikes
- • methane gas extraction
- — irrigation system

ATLAS - Part One | European Delta Landscape | Geographies and Comparison

Biographies

DANUBE
Monica Runceanu

EBRO
Lorenzo Chelleri

GUADALQUIVIR
Enrico Anguillari

NEMUNAS
Antanas Dumbrauskas
Saulius Vaikasas

PO
Maria Chiara Tosi

RHINE
Taneha Bacchin

RHONE
Fabio Vanin

DANUBE

Monica Runceanu

Intro

The largest and best preserved of Europe's deltas, situated in the south east of Romania, in Tulcea County, and generated by the Danube River, the geographic unit of the Danube Delta is defined by the three branches of the river - Chilia, Sulina and Sfântu Gheorghe and by the Black Sea shore. 82% of the deltaic surface belongs to Romania and 18% to the Ukraine.

During the 20th century the Danube Delta registered important change mainly due to the dyking of large areas for agricultural purpose, intensive fish-farming and forestry - which resulted in alterations or disturbances of the natural characteristics - and the loss of specific habitats. When the transformation measures stopped in 1990, the effects were 97,408 ha dyked area (22%), 450,000 ha of a total floodplain area of 540,000 ha cut off from the river dynamics and transformed into agricultural land, the loss of the floodplain on the Lower Danube with negative effects, embankment and separation from the river involving drainage processes - steppisation and to some extent salinisation of the soils, a greater aridity in some areas, and the cutting of artificial channels which involved changes such as the pouring of unfiltered water from the river into the inner area of the Delta. In 1990, rehabilitation/restoration studies were begun in the Danube Delta as a result of its having been declared a Biosphere Reserve. The restoration and conservation of biodiversity in a sustainable development approach became the priority tasks of the Danube Delta

Biosphere Reserve Authority

In 1991 the Danube Delta was included by UNESCO in the World Heritage list as a "Biosphere Reserve". It enjoys triple international protection status as it has also been designated a *Biosphere Reserve* by the UNESCO Committee "Man and Biosphere", and a *Wetland of International Importance* by Ramsar Convention on Wetlands Secretariat and the World Heritage site recognized by UNESCO. In 1995 the Danube Delta Biosphere Reserve was awarded the "Eurosite" prize and in 2000 the Council of Europe awarded it the European Diploma as a symbol of recognition for the environmental protection actions undertaken there.

Physical and socio-economic features of the delta

Like most delta territories, the Danube Delta is a relatively young unit and the result of the main factors governing the coastal areas, most-lyrepresented by the variation of the sea level, currents, waves, the water flow and transported alluvium, and the landscape's configuration. By considering the geographic landscape and integrating hydrographic, vegetation and hypsometric measurements, 3 units have been identified: the *Chilia – Sulina* unit (also known as the Letea unit), comprising the area covered by the Chilia branch, Tulcea and Sulina and the Black Sea shore , the *Sulina – Sfântu Gheorghe* (or Cormoran) unit, comprisingthe territory between Sulina and Sfântu Gheorghe branches and the Black Sea shore, and the Dranov unit located south of Sfântu Gheorghe branch, east of Razim Lake and the Black Sea shore. The average altitude of the Letea unit is 0.81m, 0.17m in unit Dranov and 0.37m in Caraorman, while in the Razim-Sinoe complex the average altitude is 1 m and approximately 77% of the gulf is under the sea level, since 20% of the Danube Delta is situated under the sea level. Compared with 1911 the evolution of the Delta territory indicates that approximately 1150ha/year are slipping below sea level, revealing the development trend for the entire delta area.

The delta climate is continental, influenced by the its vicinity with the sea and by the large amount of internal water areas. The average temperature is about 11°C and summer lasts 80 days in Sulina and 100 days in Tulcea, the region being at the same time the most dry in the country due to the small amount of precipitation (- 400-450 mm/year). According to resource distribution, economic and social activities and the technical infrastructure, the territory could be divided into six functional and spatial areas. The first area is insufficiently promoted for tourism due to the lack of tourism and technical infrastructure, and can be divided into the delta area, the lagoon area and the hills area, the latter being mainly a rural area. A mixed-agriculture area, consisting of large plots of land devoted to agriculture, is situated in the southern part, while another two agricultural areas are located near Tulcea Municipality and an area blending viticulture and orchards in is found in the Lunc vi a-Isaccea-Niculi el administrative unit.

An analysis of the population dynamics for the year 2007 has indicated a 3.4% decrease in the total population and a decrease of population for urban and rural areas of 3.3% and 3.7%, respectively. The population density is 30 inhabitants/km², less than the national average of 90.3 inhabitants/km² and the regional average of 79.1 inhabitants/km², mainly due to the delta specificities of the area. In urban areas the population density is 161 inhabitants/km², while in rural areas the survey registered 14.5 inhabitants/ km², the most dense area being Tulcea Municipality (521.2) while the least dense is Sulina (13.9); more than 60% of the rural areas – situated in the eastern half of the delta territory registered less than 30 inhabitants/km².

Development Issues

The development issues of the area concern the development context at the regional and national level, as well as zoning, the environment, human settlements,the population and technical infrastructure. The area faces the issue of low accessibility on land, as the Danube can be considered a natural barrier, but has a high tourism development potential, considering the influence of the neighboring development areas: Tulcea, Br ila, Gala i, Constan a. The land's main use for agriculture, forestry, reeds, fishing and tourism has meant problematically that agriculture covers an important part of the area but has low potential due to the soil's fragility (salted/ sandy soils). Opportunities are offered, however, by the possibility of using the forest for local economic development and also offered by the fact that the area for reeds and fishing covers the largest surface area, though it faces issues of increasing water pollution and its management. The three main tourism areas - DDBR, Macinului Mounatins-Isaccea-Niculitel, and the lagoon - might use the diversity of the area's natural, leisure and ethno-cultural resources in a sustainable approach, balancing the benefit of tourism flows and the effects on the ecological capacity of each ecosystem. Anthropogenic processes, resulting mainly from the large agricultural and forestry activities, though also including hydro-technical works affecting the natural flows of water and the water pollution,still threaten the environmental integrity of the area, affecting its ecological balance and biodiversity. and derive also from the The territory is a rural one: more than 90% of the settlements are rural and are confronted with isolation. Moreover, these villages are poorly connected with the districts centers, and sewage systems are lacking in some of the communes.

Policies and Planning in the Delta

Starting in 1938 when 'Letea Forest' was declared a nature reserve by the Council of Ministers, several decision enforced a regime of protection in the delta area: Rosca-Buhaiova in 1961 (protecting 14,60Oha), as well as St George-Perisor -Zatoane (16,40Oha), Periteasca-Gura Portitei (3,900ha) and Popina Island (98ha), in 1971 Caraorman Forest (840ha) and Erenciuc Forest (41ha), in 1975 the Danube Delta protected areas were extended to cover 41,500ha, in 1979 an area of 18,145ha combining Rosca-Buhaiova Reserve and Letea Forest was designated as the "Rosca-Letea Biosphere Reserve". In 1990, 500,000ha – which included all previously designatedareas - were declared a biosphere; this area was further enlarged in early 1991 to 547,000ha. International recognition came with the submission in May 1991 to UNESCO for the land to be nominated a biosphere reserve and with a similar petition to the Ramsar Bureau for the territory's nomination as a Ramsar site.

Conventions and treaties in the field of preservation and environmental protection:

- Convention (Unesco) on Wetlands of International Importance - Ramsar – 1971 (amended by the Paris Protocol, 1982)

- Man and the Biosphere Programme - UNESCO, 1971 (creation of biosphere reserves)

- Convention on the Conservation of European Wildlife and Natural Habitats in Europe – Bern

1979 (Council of Europe)

- Convention on Biological Diversity - Rio de Janeiro 1992 Jakarta Mandate (UN)

- Convention on International Trade in Endangered Species of Wild Flora and Fauna Washington 1973 - Amendments: Bonn, 1979

- The European Landscape Convention - Florence, 2000 (Council of Europe)

- Convention on the Conservation of Migratory Species and Wild Animals - Bonn, 1979

- Convention to Combat Desertification - Paris, 1994 (UN)

- Espoo Convention (1991) on EIA in the cross-border context

- Aarhus Convention (1998) Access to Information, Public Participation in Decision-making and Access to Justice in Environmental Matters

- Ramsar Convention (1971) on Wetlands of International Importance

In the program supported by the Council of Europe for the purpose of a pan-European strategy for the conservation of biological diversity and landscapes, Romania, Moldova and the Ukraine signed in 2000 the Agreement on implementation of the transboundary Reserve "Danube Delta and Lower Prut River area." In 2000, Romania together with Bulgaria, Moldova and the Ukraine created the "Danube Green Corridor" cooperation program. The Danube Delta Biosphere Reserve Authority concluded cooperation agreements with other similar wetlands, including the Memorandum of Understanding between the Danube Delta Biosphere Reserve, National Research Institute "Delta" (Romania) and Dunaisky Plavni Nature Reserve Administration (Ukraine) (1996-1999), on cooperation in personnel training, studies on biodiversity, management, and the rehabilitation of wetlands.

The International Commission for the Protection of the Danube River (ICPDR) in Vienna coordinates all activities under the Convention and is the main decision-making body of the Convention. Romania became a member of this International Commission in 1995, which was ratified by Law no. 14/1995, the Convention on Cooperation for the Protection and Sustainable Use of the Danube River. ICPDR serves as a platform for coordination to develop and initiate a basin Management Plan of the Danube river basin.

In 1999 saw the creation of the "Green Corridor" of the Danube, in a Romanian initiative, by the signing of a cooperation protocol between the Ministries of Environment of Bulgaria, Moldova, Romania and the Ukraine, which created a system of protected areas along the Danube, including the Danube Delta.

In June 2000, the Ministries of Environment of the Republic of Moldova, Romania and the Ukraine signed an agreement establishing "zones of nature protection in the Danube Delta (Delta Biosphere Reserve, Romania, Danube Biosphere Reserve – the Ukraine) and Lower Prut (Lower Prut Scientific Reserve - Moldova)".

EU Strategy for the Danube Region is a regional cooperation project which was promoted at the EU level by Romania and Austria, when the European Council of 18-19 June 2009 asked the EC to develop by the end of 2010 an "EU Strategy for Danube Region". The result is the Communication on the EU Strategy for the Danube Region and the Action Plan for the Lower Danube Euroregion, including Tulcea, Braila and Galati counties (RO), Cahul and Cantemir (MD), and Odessa (UA).

Conclusion

The Action Plan Lower for the Danube Euroregion (2009-2010) emphasizes an integrated place-based approach, "good links between urban and rural areas, fair access to infrastructures and services, and comparable living conditions to promote territorial cohesion, now an explicit EU objective".

Within the actual national planning framework, the *Zonal Territorial Plan (ZTP) and Spatial Development Strategy for Danube Delta* plays the role of the main instrument used to guide the spatial organization and to formulate spatial development solutions for the specific issues within the area. While the ZTP for the Danube Delta concerns an area of approximately 724,000 ha, it involves more than the geographic limits of the Danube Delta Biosphere Reserve (DDBR) and includes the protected area (DDBR), the municipalities included in the protected perimeter and the municipalities situated in the area of influence. The general development concept was strongly influenced by the importance of the natural environment, which in turn has been enforced by the Biosphere Reserve statute. Strategic directions and planning measures proposed by the plan are designed to work within complex relations between the natural and anthropic environments, trying to balance and resize relevant dynamic spatial processes.

The main requirements formulated and addressed by the plan concern improving accessibility in the area, developing port infrastructure, superstructure and waterway infrastructure, planning the territorial infrastructure development, supporting economic and social development, the conservation and restoration of the natural environment, identity preservation and promotion of cultural heritage and supporting rural development measures.

Development principles have been subordinated to a sustainable approach as follows: restoration, improvement, conservation of biodiversity of natural ecosystems, protection, and conservation of the natural and built heritage, preservation and increase of the area's identity; socio-economically balanced development to ensure improved living conditions; superior capitalization of natural resources; increasing competitiveness in the area within the county and regional economy; conservation of land and water areas, combat natural hazards; development of technical infrastructure requirements related to network locations and environmental limits.

The development options were related to the exploitation of existing natural resources, especially fisheries, agriculture, forestry, tourism and water; conservation, rehabilitation and protection of natural and built heritage; restructuring and modernization of technical infrastructure, especially in rural areas; balanced development of the network of existing settlements, in conjunction with the regional network of locations to ensure better living conditions; directing the expansion of settlements in accordance with local interests; equal economic and social development in territorial, public interventions and public/private, in disadvantaged areas.

Within the Spatial Development Strategy formulated for the area the measures included the institutional responsibilities and the implementation stages; integrating the documents already in force for the delta area, dedicated to the main spatial structure elements and subordinated to the main development objective *'Conservation and protection of environment achieved in harmony with a sustainable and balanced social economic development, based on the efficient use of local resources, settlements and infrastructure modernization, which have the effect of improving the quality of life in the area'.*

References

- European Union Strategy for Danube Region EUROPEAN COMMISSION, Brussels, 08/12/20010, COM(2010) 715, COMMUNICATION FROM THE COMMISSION TO THE EUROPEAN PARLIAMENT, THE COUNCIL, THE EUROPEAN ECONOMIC AND SOCIAL COMMITTEE AND THE COMMITTEE OF THE REGIONS, (SEC(2010) 1489), (SEC(2010) 1490), (SEC(2010) 1491)

- Studies and researches on the national and regional spatial planning systems, University of Architecture and Urban planning Ion Mincu (UAUIM) Bucharest and University of Bucharest CICADIT for the Ministry of Development, Regional Development and Tourism, 2009, UAUIM Team: Lect. Liviu Ianasi, Prof. Doina Cristea, Lect. Gabriel Pascariu, Lect. Claudiu Runceanu, Assist. Liviu Veluda, Assist. Adrian Cioangher and Associate Prof. Monica Radulescu as Project Director, CICADIT University of Bucharest coordinated by Lect. geogr. Daniela Zamfir.

- Planul de Amenajare a Teritoriului Zonal Delta Dun rii, Strategia de amenajare a teritoriului, INSTITUTUL NATIONAL DE CERCETARE – DEZVOLTARE PENTRU URBANISM I AMENAJAREA TERITORIULUI URBANPROIECT – BUCURESTI, Sef project arch. Constantin Chifelea Beneficiar: Ministerul Dezvolt rii Regionale di Locuin ei, 2009

- *Ecological restoration in the Danube Delta Biosphere Reserve/Romania,* WWF Germany, *Chair WWF- Institute for Floodplains Ecology, Department of Water and River Basin Management University of Karlsruhe, Danube Delta National Institute for Research and Development DDNI Tulcea, România, Kraft-Druck Ettlingen, 2008, ISBN: 978-3-00-025585-4*
- *Institutul National de Cercetare Dezvoltare în Turism, arch. Victor TIMOTIN, ec. Alina CÂRLOGEA, ec. Alina Cristina NICULESCU, ec. Doru TUDOR-ACHE, Daniela TEFAN – tehnician, Studiu de Turism pentru elaborarea Planului de Amenajare a teritoriului Zonal Interjudetean rezervatia Biosferei delta Dunarii, Bucuresti 2008*
- *http://whc.unesco.org/en/list/588*

EBRO

Lorenzo Chelleri

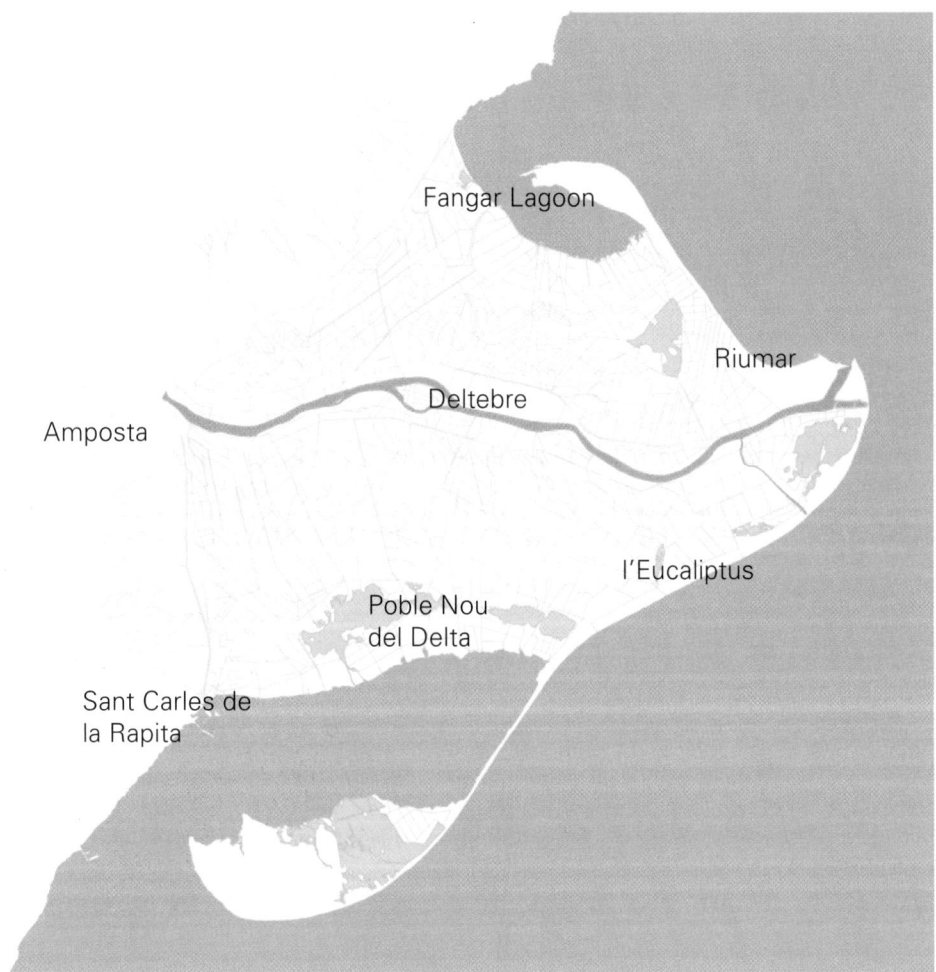

Intro

Deltas host around three-quarters of the global population, although the Spanish Ebro delta seems to dispute this idea, being a rural delta inhabited by few people. Nevertheless, it represents a significant place for the economy (thanks to its paddy fields), as a result of the high-level engineering management undertaken in the last century, which maximized potential profits from the delta's abundant environmental services. These services, distributed throughout the 320 Km^2 of the deltaic plain, today represent the biological heritage of 6-7,000 years of landscape evolution, which began with the last relative stabilization of the sea level (Mikhailova et al, 2003). Though wind, waves and river sediments are the main driving forces in a delta's evolution, in the Ebro Delta, as in most rural deltas, the anthropogenic drivers of change play a stronger role in shaping the delta's dynamics, stemming from a desire to maintain stable social-ecological (usually productive) configurations (Valdelmoro et al, 2007). As we will see, the consequence of this social (market-oriented) view is that today natural dynamics are seen more as threatening agents (to the socio-economic configurations) than as essential components of the landscape. To sustain such an hypothesis, the chapter is divided into three main parts: the first introduces geomorphologic delta features, while the second explores the delta's vulnerabilities; it is only in the last part that we will enter into landscape planning and management.

Physical and socio-economic features of the delta

As previously mentioned, the shape of the Ebro Delta has been highly dynamic and influenced both by natural and anthropogenic forces throughout its history (Sanchez-Archilla et al, 1997). Sediments supplied by the Ebro river and the intensity and frequency of winds have resulted in the formation (and evolution) of two long lobes and sandbars, such as the *Trabucador* (Southern lobe) or huge sandbanks such as the *Punta del Fangar* (Northern lobe) with the respective marine bays.

The mean river discharges have varied considerably in recent decades. In fact, climatic factors have resulted in a range from less than 50m^3/s during dry periods to more than 10,000 m^3/s during the periods of the greatest flooding (Guillén and Palanques, 1997). Most of the average changes (and consequent dramatic decreases to the sediment supply) are the result of numerous dams built since the 60s along the Ebro river. While most of the Spanish coastal areas started to suffer from speculative and massive urbanization in the 50s, this delta has most felt the effects of canalization and agricultural conquest, while urbanization has remained a lesser driver of change and pressures. In fact, the total delta population today numbers only 62,766 inhabitants (*Institut d'Estadística de Catalunya*, IDESCAT 2010), distributed within the two major towns of Deltebre and Sant Jaume (located at the centre of the delta), two small rural towns (Poblenou del Delta and Els Muntells) and two coastal tourism developments (Riumar and Els Eucaliptus). Other towns are located at the limits of the delta, like l'Ampolla or Sant Carles de la Ràpita. Recently, tourism has represented an important source of income for these coastal towns, though historically the delta-based ecosystem services were the strategic territorial aims behind locating those urban areas there. Since wetlands and lagoons are the richest and most important places for biodiversity and ecological productivity, we have to highlight the negative impact of various economic facilities and uses, like hunting and fishing, fertile soils for agriculture or for the harvesting of reeds and rushes that the Ebro delta has traditionally provided (which are nowadays regulated by the natural park). Apart from the paddy fields that occupy most of the delta's land surface, its coastal area is rich in lagoons and salt marshes (natural grasslands dominated by *halophytes*, plants adapted to waters of high-salinity just behind the dunes. Some of those marshes were transformed for use by the salt industry, which gained space among the various uses of the deltaic land, until last century, when more market-convenient rice paddies slowly substituted them. Today, one large, important salt pans is still present, located at the southern Punta de la Banya. All access to and supply by the ecosystem services (Brenner, 2007) is now regulated by the nature park, instituted in 1983.

Fishing is also a fundamental economic activity providing approximately 15% of the Catalonian annual production, though fishing in the coastal lagoons has recently been reduced due to over-exploitation. In fact, while wetlands and lagoons represent the most species-rich habitats, today these represent only 25% of the natural deltaic landscape (*Parc Natural Delta del Ebre*, 2005) due to the delta's exploitation for agriculture, which has gradually reclaimed nearly all the fertile land available in the delta (Sánchez-Arcilla et al, 1997). Within this 25%, the surviving ponds, which have soils too salty for exploitation as rice paddies, have their origin in the isolation of a former sea bay with sand bars. The biological richness of the coastal lagoons stands in sharp contrast to the intensive rice agricultural use dominating the entire delta landscape, turning the delta into a never-ending swath of green, blue or brown, depending on the season, as the rice cycle's crop must be flooded and dried cyclically during the year. In 2012 those paddies cover almost 65% of the delta's surface, representing 98% of the total production of this cereal in Catalonia and making it the third-largest source of rice on the European market thanks to the 120,000 Mt/y (metric tons per year) it produces (Day et al, 2006).

What is socially and ecologically interesting about the rice fields is that they act as a wetland for half the year (when they are flooded). In so doing, these (freshwater) flooded lands provide important ecological functions, serving as temporary wetlands (a strategically important habitat for migrant birds species), and regulating the intrusion of saltwater into the ground, as this temporary freshwater table keeps the saltwater downstream (Sánchez-Arcilla et al, 2008). For this reason, notwithstanding the intensive agriculturally-based use of the delta, paddies ably balance the economic and environmental needs of the delta landscape.

The Social-Ecological Vulnerability of the Ebro Delta

Our present society has underestimated the environmental services (henceforth, ES) supporting our well-being; indeed, only in this last decade has there begun to be a proper study and consideration of their value (Costanza et al, 1997). Deltas are living examples of the most productive ecosystems (just think that much of the marine fishing worldwide is associated with some form of delta-provided ES) and despite their sensitivity to external environmental changes we know these territories are highly resilient, thanks to the natural interplay between constructive and destructive forces (Vorosmarty et al, 2009). Unfortunately, anthropogenic pressures and influences have been decreasing the delta's resilience through command and control practices that seek to stabilize environmental services by constructing dams at upstream drainage basins and through canalization of the downstream river deltas. As hydraulic engineering is now a pre-dominant global force, increasing reservoir construction by 600–700%, the sediment flows have tripled in the time needed to reach the sea and thus to help coastal areas facing (increasing) erosion patterns. The key role of sediment flow lies in its capacity to make "delta plains survive a high rate of eustatic Sea Level Rise" (SLR henceforth) (Day et al, 2011:490). From the IPCC's most recent predictions, SLR alone will increase deltas' vulnerable areas by 50% by 2100 (Overeem and Syvitski, 2009). However, despite CC variables evidencing an increase of precipitation rates over the last 10 years, a reduction can be observed in the average annual flow of the Ebro river: an 11% decrease from 1913 to 1935 and in 1951 and 1979, a 23% decrease from 1951 to 1970 and a 19% decrease from 1971 to 1990 and in 1991 and 2004 (Fatoric and Chelleri, 2012). This reduction is essentially to be attributed to the construction of dams along the upstream river basin (Mequinença, Flix and Riba-roja) from the 1960s onwards.

The interplay of forces both constructive (sediment flows) and destructive (storm-wind actions) for centuries kept the delta in a state of high ecological productivity and maintained plain accretion. However, as Vörösmarty et al. states, "there is a basic tension between the human desire for stability and the dynamism by which natural deltas maintain themselves" (Vörösmarty et al, 2009:33). From such tensions the vulnerability of the Ebro Delta began to rise in proportion to the number of canals being built. From 1900 to 1970, most of its wetlands and lagoons were transformed into rice paddies, and an intensive drainage system was built to supply them with river water. As high floods were still present in the delta, the construction of dams aimed to regulate the flow of water and to maximize rice production. This resulted in a drastic reduction of sediment transportation, reducing by 99% the level of sediment (50cm) deposited between 1860 and 1960 (Sánchez-Arcilla et al, 2008). Present sedimentation rates range from 4 mm/year in the wetlands at the river mouth to less than 0.1 mm per year in impounded salt marshes. This delta thus passed from a vertical accretion of 0.5cm per year to a net sediment loss from the delta plain, intensifying the subsidence process already currently estimated at around 2-5 mm/year (Day et al, 2006). Furthermore, since roughly 40% of the emerged delta plain has an elevation of just less than 50 cm (and about 10% is already below sea level), SLR represents a real threat in the middle- and long-term. All these assessments of the various instances of environmental impact (*Taller d'Enginyeria Ambiental*, 2008) sum up the vulnerability of the Ebro Delta, which needs middle-term (socio-economic) transition strategies to face the future environmental challenges.

Policies and Planning in the Delta

The numbers from the Spanish Millennium Assessment reflect the fact that in the last three decades Spain's greatest economic growth has come from urban sprawl. In fact, from the 80s, until now, coastal areas have suffered an increasingly unsustainable pressure from human development, and almost 60% of the total wetland area has been lost. Just 20% of coastal dune systems have been preserved in a positive ecological state and 70% of the coastal lagoons have disappeared as a result of urban development (Barragán and Borja, 2010). In 2009 more than 40% of the total Spanish population lived along the coastal area, to which were added around 45 million tourists during summer. This pressure has resulted in almost 60% (13 of 21) of the Ecosystem Services being dramatically depredated or subjected to unsustainable exploitation (Barragán and Borja, 2010).

In the 80s the Ebro Delta, as part of the Spanish coastal fringe, saw urban sprawl plans reach its lagoons. However, interests in its strong rice economy could put a halt to any urban sprawl project; moreover, some environmental protection laws have begun to be implemented in the territory. In fact, the Ebro Delta Natural Park (Decree 357/1983, extended by Decree 332/1986) was created by the regional government in 1983. This initially covered just 8,445 ha on land and 564 ha in the sea; today, thanks to "Natura 2000", a European network that establishes Special Protection Areas for wild birds, and other environmental protection policies, it has reached a total area of 12,738 ha inland and 35,647 ha in the sea. Though included in the RAMSAR list of Wetlands of International Importance in 1993, it was not until 1995 that urban and territorial planning started to regulate land use in the delta region.

The first Ebro Delta Development Plan (*Plan Director del Delta del Ebro*) in 1995, and the Territorial Plan of the Ebro Region (*Pla Territorial de les Terres de l'Ebre*) of 2001, were developed in a poorly integrated framework, and while water management was entrusted to the Hydrological Plan (NHP), land use was regulated by these two plans aimed at developing tourism services, infrastructures and industrial activities in the delta region (mainly along the north-south axis, linking Alicante and Valencia with Barcelona). Since the water management practices proposed by the 2001 NHP, which should have followed the UE Water Framework Directive (WFD) in a sustainable and integrated manner, were accused of and rejected for employing dramatic unsustainable principles (Day et al, 2006), a new integrated framework for spatial planning was proposed for the Ebro Delta region (Ministerio de Medio Ambiente, 2007) within the new Ebro region territorial plan (*Pla Territorial de les Terres de l'Ebre, 2009)* and the Integrated Plan for the Ebro Delta Protection. In the meantime, a

reformulation of the national hydrologic plan was supposed to have occurred.
Because the annual turnover produced by this delta's ecosystem services stands at almost 120 million (from fisheries, aquaculture, agriculture and tourism), the new integrated approaches undertaken by the territorial and delta protection plans involve: defining hydro regimes in order to maintain the basic ecological functions (the rice paddies lobby is a powerful player in pushing for water availability, prioritizing the agriculture sector's water demand over that of other uses), restoring cannels and improving the condition of habitats, building a strong monitoring system of the environmental indicators, improving the quality of water for agriculture and avoiding urban sprawl (only very limited and compact growth is planned) througout the delta coastal region (*Departament de Política Territorial i Obres Públiques*, 2010). These plans attempt once more to stress firmly something that since 1988 with the Coasts Law (Ley de Costas, *22/1988*), and afterwards in 2004 with the Coastal Directors Plans (*Pla Director Urbanístic del Sistema Costaner - PDUSC*), has been quietly neglected in Spain: coastal protection against urban sprawl and land use speculation.

Conclusion

As reported in the various sections above, the Ebro Delta as a social-ecological system is under the threat of CC (and its socio-economic regime). The decrease in the delta's resiliency highlights its heavy usage as a support for the rice market. It seems rice lobbies are constantly negotiating for sufficient freshwater for rice paddies, versus other actors claiming more water for maintaining delta's coastal ecosystems. However, the emerging planning and environmental management policies and practices are trying to address short-term challenges (such as avoiding construction, improving channel efficiency and building the environmental monitoring system) while more integrated strategies addressing longer-term challenges are hopefully next to de debated and proposed for the Ebro Delta.

References

- Barragán, J.M., Borja, F., (2010) *Evaluación de los ecosistemas del Milenio de España*. Litorales. Madrid, Ministerio de Medio Ambiente, Medio Rural y Marino. pp. 673-739
- Brenner, J., (2007) *Valuation of ecosystem services in the Catalan coastal zone*. SARDA, R., Jiménez, J., (directors). Doctora thesis, Universitat Politècnica de Catalunya, Departament d'Enginyeria Hidràulica, Marítima i Ambiental
- Cardoch, L., Day, J.W., (2002) "Biophysical energy analyses of non-market values of the Ebro Delta", *Journal of Coastal Conservation*, 8: 87-96
- Costanza, R., d'Arge,R., de Groot, R., Farber, S., Grasso, M., Hannon, B., Limbert, K., Naeem, S., O'Neill, R. V., Paruelo, J., Raskin, R. G., Sutton P., and van den Belt, M., (1997) "The value of the world's ecosystem services and natural capital" *Nature* 387: 253–260
- Chelleri, L.; Schuetze, T.; Ridolfi, E.; Trujillo, A. and Breton, F. (2011) "*Navigating European Deltas' vulnerabilities and resilience: learning from different adaptation approaches and experiences how to cope with changes in social ecological systems*" Poster Presentation at the Second Conference Resilience 2011, Arizona State University, Tempe, Phoenix
- Day, J.W., Maltby, E., Ibañez, C., (2006) "River basin management and delta sustainability: a commentary on the Ebro delta and the Spanish national hydrological plan", *Ecological Engineering* 26: 85-99
- Day, J., Ibáñez, C., Scarton, F., Pont D., Hense, P., Day J. and Lane R. (2011) "Sustainability of Mediterranean Deltaic and Lagoon Wetlands with Sea-Level Rise: The Importance of River Input", *Estuaries and Coasts* (34):483–493
- Fatoric, S., and Chelleri, L., (2012) "Vulnerability to the effects of climate change and adaptation: the case of the Spanish Ebro delta", *Ocean and Coastal Management Journal* (60): 1-10.
- Guillén, J. and Palanques, A. (1997) "A historical perspective of the morphological evolution in the lower Ebro river", *Environmental Geology* 30

(3/4): 174-180
- Institut d'Estadística de Catalunya (2010) *Anuari estadístic de Catalunya 2010.* http://www.idescat.cat (accessed April 1, 2010).
- Mikhailova, M.V., (2003) "Transformation of the Ebro river delta under the impact of intense human-induced reduction of sediment runoff" *Water Resources* 30 (4): 370-378
- Overeem, I., and Syvitski, J.P.M., (2009) *Dynamics and Vulnerability of Delta Systems.* LOICZ Reports & Studies No. 35 (Geesthacht, GKSS Research Center).
- Sánchez-Arcilla, A ., Jiménez, J.A ., Gelonch, G., Nieto Romeral, J. (1997), "El problema erosivo en el delta del Ebro", *Revista de obras públicas* 3368, 23-32
- Sánchez-Arcilla, A ., Jiménez, J.A ., Valdemoro, H.I., Gracia, V., (2008) "Implications of climatic change on Spanish Mediterranean low-lying coasts: the Ebro delta case", *Journal of Coastal Research* 24 (2), 306-316
- Taller d'Enginyeria Ambiental (2008) *Estudis de base per a una estratègia de prevenció i d'adaptació al canvi climàtic N1: delta de l'Ebre.* Document de síntesi. Barcelona
- Valdemoro, H.; Sánchez-Arcilla and Jiménez, J. (2007) "Coastal dynamics and wetland stability. The Ebro delta case", *Hydrobiologia*, 577, pp. 17-29
- Vörösmarty C.J. et al., (2003) "Anthropogenic Sediment Retention: Major Global-Scale Impact from the Population of Registered Impoundments," *Global and Planetary Change*, vol. 39, pp. 169–90.
- Vörösmarty C., Syvitski, J., Day, A., Sherbinin, L., Giosan L and Paola, C., (2009) "Battling to save world's river deltas", *Bulletin of the Atomic Scientists* 65: 31-43

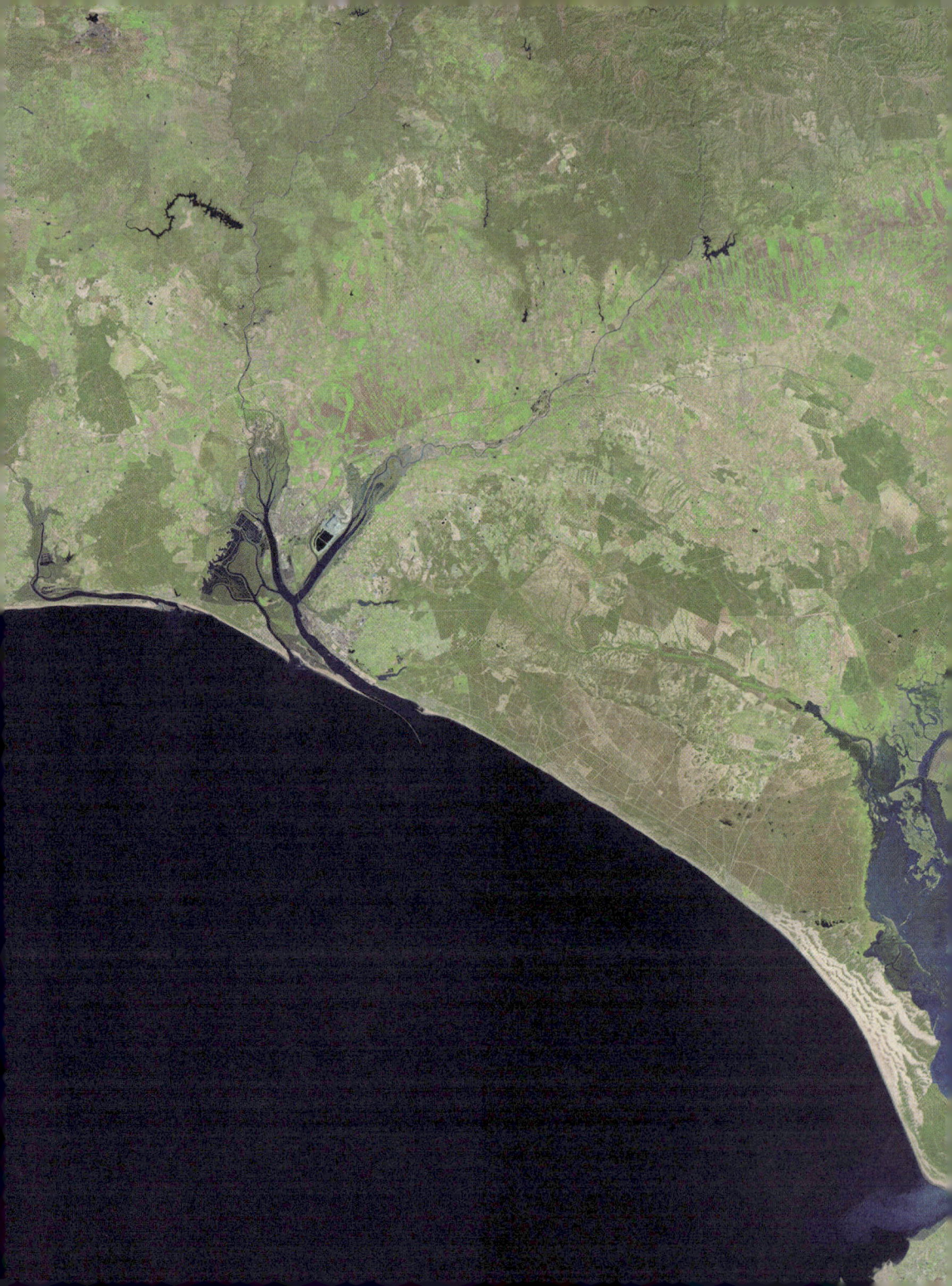

GUADALQUIVIR

Enrico Anguillari

Notes on transformations, conflicts and planning for the Guadalquivir estuary

The Guadalquivir flows through the western part of the Mediterranean biogeographical region, at the junction of two continents and two large bodies of water with very different characteristics. It passes through the entire width of Andalusia, encountering the highest mountains of the Iberian peninsula, covered with snow for most of the year, and areas among the wettest in Spain and large desert regions. It is born in the Mediterranean region and reaches the ocean in the Doñana area.

Here, its estuary moves inside the triangle formed by the cities of Huelva, Seville and Cadiz, over an area of about 290,000 hectares hosting a population of 175,000 inhabitants, distributed in 14 municipalities, though more than two million people live in the immediate surroundings. Thus, the area of Doñana finds itself at the centre of a large metropolitan context, whose agglomeration thins out towards the 'comarca' of Campo de Gibraltar and along the Bay of Cadiz, assuming the characteristics of the low-density urbanization, urban dynamics and mobility, and level of land use typical of urban sprawl.

The Doñana is one of the most important and unique natural areas in Europe and one of the main stops in the migration routes of many species of birds between Africa and Europe. The importance of this area is confirmed by many national and international instruments protecting its specificity. In addition, the Doñana can be considered a cultural landscape typical of the Mediterranean.

While its central part is strictly protected, the surrounding area is almost entirely used for rice cultivation. The coast is instead linked to tourism and displays an urbanization that alternates seaside resorts with a more dispersed and fragmented fabric of second homes.

From the middle of the last century, the Doñana area has been the subject to major transformation policies, halted somewhat by the establishment of a National Park in 1969. Its dichotomy as a protected area/production space is still the cause of violent conflicts and, on the part of the local population, of a general feeling of rejection towards the Park for fear of being 'excluded' from development opportunities. Consequently, although there has been a general improvement in the management and protection of the area, the measures adopted have not yet been able to stop the many pressures threatening the entire region.

Changes

Like much of the alluvial territories, the Guadalquivir estuary has been modified by the joint action of man and nature.

Until the sixteenth century, the current region of Doñana was a wide gulf - the Tartésico Gulf - which made Seville one of the most important ports for trade with the Americas. In the space of two hundred years, the sediments carried by the river have gradually buried the gulf, severing its dependence on the sea entirely and transforming it into a swamp river, or rather, into an 'inland delta' consisting of three branches surrounding two large islands: the Isla Mayor and the Isla Menor. Due to the progressive reduction of the seabed, the Guadalquivir has not been navigable since 1680; as a result, the fleet of the Indies and of the *Casa de Contratación* were transferred from Seville to the harbour of Cadiz. From here, there began a 'season' of large engineering projects aimed at making the river navigable again and protecting the population from floods that periodically struck the city of Seville. This involved a long series of corrective interventions that, from the digging of the Merlina in 1795, lasted until the excavation of the Chapina canal, for the 1992 Expo. The works - ten in all - have reduced the river's length by 50 km and at the same time have caused a significant reduction of the flow of the river's lateral branches. Along with these corrective interventions, 33 basins were added built along the main watercourse and at the convergence of the river with its tributaries, to protect Seville from floods. Today, almost all the branches of the Guadalquivir are rigidly channeled through embankments as much as 12 meters high. The effects of these interventions have been felt throughout the entire territory. The concentration of the flow along the main shaft and the canalization of the estuary have led to the reduction of seasonal flooding in much of the river's floodplain blocking the natural process of sedimentation and changing its ecological balance.

In turn, this has led to the beginning of the gradual reclamation of the land and its use for human activities. In the nineteenth century, a plan was made to drain the marshes for breeding cattle. However, the work did not begin until the 1930s, when the Marismas became the object of strong speculative interest by two large foreign companies (Drain, 2003). With British and Swiss capital, in 1926, the *Compañía de las Islas del Guadalquivir* was created, which, after having bought the Isla Mayor and 23,000 hectares on the right side of the river, began a rapid process of reclamation which, only three years later, allowed the first crops of rice to be planted. The works transformed the top half of the island into a 14,000 hectare polder, with 45 km of drainage channels and two dewatering pumps, linked to numerous works and infrastructure for the development of the entire region - 68 km of roads, 54 railways, 60 phone lines, 27 power lines and 5 villages for the accommodation of the settlers. The second reclamation campaign was begun in 1928 by the *Compañía de las Marismas*, established in 1921 with French capital. The works provided for the construction of four large polders. The first two, of 7,000 hectares, were completed in 1930 and 1931. The third, of 14,000 hectares, was completed in 1934 and the fourth, closer to the sea, on more saline soils, was never finished. The political and social tensions which led to the civil war brought both companies to sell and abandon the land, until 1938, when, in order to meet the food demand of the republican troops and nationalist Spain, the cultivation of rice was taken up again, and with it, the alterations of the territory. In 1939, the *Instituto Nacional de Colonización* was created as an 'instrument' of Francoist agrarian reform policy. The following year, the INC began studying the desalination of the whole left side of the Marismas. The results of this survey gave rise to a long legislative process that, in 1955, led to the declaration of the entire area as an Area of National Interest for its potential profitability and, in 1960, resulted in its inclusion in the state programming of irrigable areas (*'Grandes Zonas Regables'*). The work of land reclamation began the same year and involved 72,000 hectares, of which 20,000 were irrigated with water taken directly from the Guadalquivir and the rest from the Bajo Guadalquivir channel. These works led to a profound change in the landscape of the marshes whose meanders have been replaced by a perfect division into plots of 500 by 2,000 meters cultivated by roughly 4,000 new settlers from a dozen newly founded villages.

While work on the left side of the Marismas were carried out without any major conflicts, the right side was declared National Area of Interest and a reclamation plan for it was approved - *Plan Almonte-Marisma* - in 1971, almost simultaneous with the establishment of the *Parque Nacional de Doñana*, generating strong conflicts between agricultural development policies and policies for the protection of the area. The object of the dispute concerned mainly, and still concerns, the negative effects of 'hard' and highly polluting agriculture that, using the ground water, places the Doñana nature reserve at risk.

Since then, there have been many studies both on the reclamation's environmental impact and its sustainability, bringing the Ministry of Agriculture - which directed, until two decades ago, both the *Instituto para la Reforma y el Desarrollo Agrario IRYDA*, the promoter of the reclamation works, and the *Instituto Nacional para la Conservación de la Naturaleza ICON* that managed the assets of the Doñana - to take an ambiguous position and promote both the transformation of the territory and the protecting of its environmental values (Ministerio de Agricultura, 1984).

From the territorial project to landscape politics

If the left side of the Guadalquivir Marismas has been irreversibly transformed, on the right side interests are focused on protecting its natural resources.

In principle, the attention which led to establishment of the *Parque Nacional de Doñana* found no significant response from institutions, which were more sensitive to the interests of the agricultural sector or those of the booming tourism sector along the coast. Rather, the Park has its origins in an initiative of a private nature. The first 10,000 acres were acquired by the WWF and sold to the Spanish Government on the condition that they be declared a nature reserve - the *Reserva y Estación Biológica*. From that moment, a hard, three-way fight began over land use, which still today has ramifications in the strategies, policies and choices for this area. On the one hand, there is the nature reserve desired by the WWF, on the other, there are tourism interests that, in 1965, led to the drafting and approval of the coastal *Plan de Ordenación Turistíca*, and thirdly, the IRYDA, backed by the FAO, which supports the use of the left part of the Marisma for agriculture. In compliance with each party's needs, in 1969 the coast of Matalascañas was declared a National Centre of Tourism Interest; in the same year the *Parque Nacional de Doñana* was established, and, in 1971, the actuation of the Plan Almonte-Marisma became a National Priority. Today, the main problems relating to the protection of natural areas of the Doñana, are the result of this paradox. First, much of the land included within the Park is private property. Second, despite the fact that the Plan Almonte-Marisma is for various reasons 'frozen', a change in the political and social conditions could at any moment revive its implementation. Finally, the development of tourism along the coast is seriously endangering the entire environmental system of the region.

Despite these tensions, the Andalusian planning system puts the landscape and environmental values at the forefront of the 'issues' aimed at guiding the overall transformations of the Doñana territory. The *Plan de Ordenación del Territorio de Andalucía - POTA*, which entered into force in 2007, aims to bring the common paradigm of 'landscape' into all territorial policies of the autonomous regional government of Andalusia (Junta de Andalucía, 2007, 2012). In this sense, it should be seen as an innovative plan for implementing the European Landscape Convention (Porcel, Hildenbrand, 2012), because it aims to make landscape management policies and territorial planning policies interdependent. In the context of POTA, the area of the Doñana thus appears as a laboratory for experimenting with this complex reciprocity. The water system, the winter storms and summer drought, the irregular yearly rainfall, the sea's influence, ocean currents, winds from the east and west, the extremely flat landscape, the presence of freshwater and saltwater, the thousands of hours of sunshine a year, all make the Doñana region rich in biodiversity, where habitats and species of interest to the European community are found, such as: 15 habitats listed in Annex I of the 92/43/EEC Directive, of which 6 are priority habitats; 16 animal species listed in Annex II, including the Iberian lynx, classified as a 'critically endangered species' by the International Union for Conservation of Nature (IUCN), 300 bird species of which 10 are included in Annex I of the 79/409/CEE Directive, and several species of plants. The Doñana's protected area comprises about 110,000 hectares, split almost evenly by the National Park and the Nature Park. Since 2007, the two parks make up the *Espacio Natural de Doñana*, aimed at the integrated management of a territory has been declared a Biosphere Reserve, a World Heritage Area, a Ramsar Area, a Special Protection Area, that is part of the Natura 2000 network and that has obtained the European Diploma for Protected Areas.

Today, the importance of the natural values represented by the Doñana and the gradual perception of its uniqueness guide a great deal of planning measures, influencing its territorial structure and its land uses. The trend is that of a gradual transition from the protection of an area considered a watertight compartment, to the management of an increasingly dynamic context, which is trying to hold together economic development and environmental conservation (WWF, 2010).

This shift can be observed by retracing the history of the laws affecting this area. In 1969 the National Park of the Doñana was created. The law constituting it established the necessity to develop a master plan for the use and management of the

National Park and a territorial coordination plan. The latter was approved in 1988, with the aim of providing the basic guidelines for the organization of the territory and a framework for the development and coordination of the policies, plans, projects and programs of local bodies and of individuals. Later, in 1989, the Doñana Nature Park was founded, consisting of a series of areas near the National Park. Here the natural resources management plan overrides all other planning tools on both the territorial and sectoral level.

1992 was the turning point, when the *Junta* or autonomous regional government of Andalusia appointed an international commission of experts coordinated by Manuel Castells to create a strategy for socio-economic development of the area of the Doñana. The results of this study were the starting point for the revision in 1993 of the *Plan Director Territorial de Coordinación de Doñana y su EntornoPlan* the *Plan de Desarrollo Sostenible para el Entorno de Doñana 1993-2000*, the *Plan de Ordenación del Territorio del Ámbito de Doñana*, approved in 2003 with the aim of defining a development model for the region and for the *II Plan de Desarrollo Sostenible de Doñana* approved in 2010.

Fragility and threats

The case of the Doñana is paradigmatic for how anthropogenic pressures have affected the hydrogeological layout and the environmental, altering them radically.

In the last century, this area has been subjected to an intense campaign of public and private interventions as part of a development model based on a profound transformation of the territory.Today, despite the cumbersome presence of the two parks and of a planning system that places an emphasis on the themes of landscape, the Guadalquivir estuary is displaying criticalities concerning hydraulics, morphology, ecology and pollution, and its balance is rendered unstable by a number of problems directly related to the cultivation of rice and the substantial tourism development which, since the end of 1900s, has been in full swing along the coast. To make room for agriculture, from the 1930s to today, 84,000 acres of marshes have been reclaimed, or 62% of the total land area of the Guadalquivir Marismas; in addition, almost 3,000 hectares of forest were cleared, including most of the riverside woods which occurred with the correction of the waterways. Currently this area boasts 40% of all the rice fields in Spain. The combined effect of the massive consumption of fresh water for the rice fields, the progressively reduced flow of the river and the 3,000 hours of sunshine a year to which the region is exposed, is gradually drying up large portions of the Doñana. Over the past 80 years, more than 100 species of plants that colonized the wetlands have disappeared, and the phenomenon is likely to worsen given that at this latitude, by 2050, there will be an increase in temperature between 1.4° C and 3.8° C, and a reduction in annual rainfall between 5 and 10% (Lop, Hernández, 2009). The disappearance of plant species combined with the deforestation of large areas has resulted in serious consequences such as increasing erosion, the reduction of the barrier effect and the loss of specific habitats. Runoff of fertilizers and pesticides used in agriculture has jeopardized the groundwater, the rivers and also the sea, causing eutrophication. This has resulted in an increase in the water's turbidity and salinity, with serious socio-economic difficulties for the rice industry, fishing, tourism and the environment.

To this are added the risks deriving from the rising mean sea level. The Doñana territory is nearly flat and its highest peaks do not exceed 40 meters. Over the course of the last century, there has been a sea eustatism of about 20 cm and, over the next 100 years, the scenarios constructed on the basis of climate change speak of an increase that could reach 110 cm (Bardají et al., 2009). This would place the remaining wetlands and their associated habitats in serious danger as, with the intensification of sea storms, they would become more and more like lagoons. The coast itself could not protect the territory behind it. For some time, studies on the movements of the coastline of the Gulf of Cadiz, based on both aerial photogrammetry and on the continuous monitoring of the profiles of its beaches, have shown that the coast is eroding (Del Río et al.

2002). The evidence of this process are the watch towers built along the coast in the late sixteenth century that show that the coastline has retreated along the beaches of El Asperillo, Sanlucar de Barrameda, and Chipiona over the last fifty years. The natural movements of the coast, which also include erosion, are mainly due to waves generated by the winds that, blowing predominantly from the southwest, cause a general drift towards the east. In front of the city of Huelva, in the last 50 years there has been a continuous advancement of the coast and long sandbars have arisen in a crest like formation. The winds from the east instead cause strong sea storms and the substantial erosion of Cadiz's coast.

The entire Gulf is 'unbalanced' also due to the high concentration of infrastructure that conditions the existing natural forces. On the one hand, the numerous works - mainly dams - built along the banks of the Guadalquivir and its tributaries have reduced sediment deposits downstream, actually giving life to the coast. On the other hand, to allow the development of tourism, construction has been dense in some parts of the coast and there its dynamics have been irreparably altered. For the most part, these tourist sites are located in areas designated as 'servitude' and 'influence areas' by the Spanish coastal law - *Ley de Costas* - which today are in serious danger due to the erosion to which the coastline is subjected. Urban expansions, and individual houses, dams, ports and groynes, besides having 'hardened' the system, have also limited nature's ability to recover from the sea surges and tempests that have been gradually been intensifying. These structures act as barriers preventing drift and the redeposit of sediments. On the central part of the Gulf of Cadiz, in front of the Doñana National Park, erosion is estimated to be between 1.25 and 2.2 meters a year. This is mainly due to interference from the Juan Carlos I Dike, built at the mouth of Huelva in front of Mazagon, one of the main seaside resorts in Andalusia. With its 14 km of length, this dam has accelerated the erosion of the whole central part of the coast, while favouring the coastline's accumulation to the west, at the mouth of the Odiel-Tinto, near the border with Portugal and to the southeast, at the mouth of the Guadiana, above Cadiz.

Today, the *Ley de Costas* is under revision and the current government seems to favour a less restrictive and more permissive reading of its principles, suggesting that urbanization could be a virtuous means to solve the problems of coastal erosion and environmental degradation. In contrast, many observers raise serious doubts as to this attitude, which could become a pretext to continue to build, further aggravating the already complex situation in the region of the Doñana (WWF, 2012).

References

- Bardají, T., Zazo, C., Cabero, A., Dabrio, C. J., Goy, J. L., Lario, J., Silva, P. G., (2009), *Impacto del cambio climático en el litoral*, Enseñanza de las Ciencias de la Tierra, 17.2, 141-154
- Drain, M.,(ed), (2003), *Politiques de l'eau en milieu méditerranéen. Le case de la péninsule Ibérique*, Casa de Velázquez
- Junta de Andalucía, (2007), *Plan de Ordenación del Territorio de Andalucía*, Decreto 206/2006 de Noviembre de 2006
- Junta de Andalucía, (2012), *Estrategia de Paisaje de Andalucía*
- Lop, A., Hernández, E., (2009), *Caudales ecológicos de la marisma del Parque Nacional de Doñana y su área de influencia*, WWF, España
- Ministerio de Agricultura, (1984), *Informe sobre la problemática y soluciones sobre Almonte-Marisma y Parque Nacional de Doñana*, Madrid.
- Porcel, O., Hildenbrand, A., (2012), *Working Landscapes 1. Landscape Strategies in Spain: a compared analysis*, Florence, RECEP-ENELC
- WWF/Adena, (2010), *Un futuro para Doñana*, Madrid, España
- WWF, (2012), *WWF rechaza la reforma de la Ley de Costas*, in: wwf.es, http://www.wwf.es/?22520/WWF-rechaza-la-reforma-de-la-Ley-de-Costas (last checked: November 2012)

NEMUNAS

Antanas Dumbrauskas ❀ Saulius Vaikasas

Physical-geographical features of the delta Nemunas basin

Nemunas is a low land transboundary river and the largest in Lithuania, with a catchment area of 97923.8 km^2. It rises in Belorussia at the confluence of two rivers, the Usa and the Losha, and flows to the Curonian Lagoon (Baltic Sea). Nemunas catchment area covers 75% of the Lithuanian territory. The river's length is 937 km:359 km of it is located in Lithuania, 462 km in Belorussia and 166 km constitutes the boundary with Russia. Lithuania contains the largest part of the Nemunas catchment are (47.5%), the rest belongs to Belorussia (46.4%) and Poland (6.1%) (Jablonskis 1993). The main tributaries are Berezina, Shara, Neris, Nevezis, Jura.

The Nemunas River Delta, shared by Lithuania and Russia, starts 48 km from the mouth, where the Nemunas splits into the Gilija and Rusne distributaries. The Gilija turns towards the Russian side and the Rusne (about 13 km from the mouth) also splits into two main branches, the Atmata and the Skirvyte. Rusne island is located among these distributaries and is divided into several parts by several smaller watercourses. Both the island and the entire delta area are furrowed by a dense network of open channels. The delta area (sometimes called the Nemunas Lowland) is usually considered to include also the Nemunas floodplain, which starts at the mouth of the Jura River. The delta area covers about 930 km^2, of which approx. 605 km^2 is seasonally inundated (Basalykas 1961).

The delta area is not homogeneous. Irregularities of more than 100m in the delta area's surface after the Quaternary period have gradually been flattened by the accumulation of moraine deposits. The result is two geologically distinct parts of the present delta. The Western part of the Quaternary cover is very thick and overlays the eroded terrain of the ancient Cretaceous formations. Meanwhile, the Eastern part is covered just by 10-20 m of Quaternary layer, located on a still-intact elevated plateau from the Neocene period. The current surface of the delta area is mainly a flat floodplain with average height of 2-5 meters above sea level, whose surface was formed by the last icing and later reshaped by the pre-glacial lakes and the accumulation of the delta. All these processes have been accompanied by persistent tectonic sinking.

Throughout the history of its development, the Nemunas delta has been and continues to be shaped by the water flow. The dominant water flows bringing most of of the sediment into the delta area have been always those of the Nemunas and its branches; some extra sediment is borne by the Minija and Shisha distributaries. The entire basin begins and ends in the north, in the Atmata and Gilija (Matrosovka) branches, which lie 48 km south of the great deltoid plain, called the Baltic slacks area of the Nemunas delta plain. This plain was formed by centuries of river sediment.

Intensification of farming in the Nemunas river basin, deforestation, land reclamation, and increase of arable land have all caused a more intensive transportation of eroded particles, depositing the river's burden of sediment in the delta floodplains. This stimulated siltation of the river bed and caused a rising river bed elevation. Therefore many of the delta branches have been forced to change directions, some meanders have been cut off from the main stream, forming ox-bow lakes. One former bay which today has become a lake - Krok Lanka – is also is a result of sealed delta deposits. A similar fate may await Kniaupas bay. The large quantity of silt has led to intensive development of the delta front, which is continuously moving towards the Curonian lagoon at a rate of several meters per year, forming new islands and small, closed-off bodies of water.

The most important part of Nemunas delta is Rusne Island. It is surrounded by the aforementioned two main branches, Atmata and Skirvyte. The total area of the island today is about 55.6 km2. The other one is Vente Cape - a small peninsula at the eastern coast of the Curonian Lagoon and Kroku Lanka Lake.

Some historical facts about the Nemunas delta

The Nemunas delta area has a long and rich history. Over the centuries this land has been *invaded* numerous times, but its longest inhabited period was that of the Baltic tribes, the Curonians and Skalvians. It later saw an increase in the number of colonists from Western European countries. Finally, in the delta area there arise the historical region known as Lithuania Minor. This community developed their own dialect and specific lifestyle. In the nineteenth century, the Nemunas Delta became a very densely populated region. Colonies began to settle at the outskirts of the delta in the Rupkalviai, Aukštumal, Berstai and Medziokles wetlands. There, for a small amount of money, newly-arrived settlers could rent 13 hectares of swampland. These settlers dug and dried peat and transported it for sale to the surrounding cities. Zalgiris village, formerly known as colony, still to this day Bismarck colony name inhabited. The left side of the Nemunas delta belonged to the German Reich until 1945. After that date, the area was incorporated into the U.S.S.R. and continues to belong to Russia today.

In 1918, the right side was placed temporarily under French administration, and in 1923 was annexed to Lithuania. In March 1939, it was ceded to Nazi Germany; after World War II it returned to Lithuania. In 1914 the first bridge was built across the Atmata branch to th island; destroyed during World War II, it was not rebuilt only in 1970.

After World War II, the population of the Shilute region radically changed. Before the war, the Germans who lived there and part of the Lithuanians fled to the West before the approaching front; the remaining part of the area's inhabitants was exiled to Siberia. Empty homes were settled by newcomers from different parts of Lithuania and the Soviet Union. Shilute town, which hada population of 4500 in 1939, counted only 7 local citizens among its inhabitants in 1944. By 1959, the population had already grown to 8,969. From 1944-1947 Shilute again become the centre of the county, and since 1950 has been the district centre.

Over the centuries in the Nemunas River Delta area a unique group of Lithuanians has developed; these Lithuanians speak a dialect that differs from those of other parts of Lithuania; as do their their local customs, architecture and the special lifestyle that characterizes this region. Most of the inhabitants of this land cannot imagine a life not marked by the presence of water and floods. There is no one in this area who is not in some way involved in fishing. Over a long period of time, the fishermen of Nemunas delta have carved out for themselves a peculiar lifestyle, comprising work customs and tools, and fishing methods, in addition to a unique local architecture.

Human activities in the delta area

The left part of the floodplain belongs to the Kaliningrad region and the territory has been embanked by high dikes since Prussian times. Most of the wet areas lie in the Nemunas river lowland, while the delta area is on the right coast of the river. In order to control flooding in the Lithuanian part of the delta, the development of a system of polders was begun in the nineteenth century. Water from the polders is extracted with pumps. The land here is protected by two types of polder dikes: "winter" polders (or polders not flooded all year long) and "summer" polders, which protect the area only during the period for vegetation (summertime). Since 1907 39 polders have been built, including 12 winter polders. The type of polder depends on the agricultural importance in the area and its flood protection policy. Almost all winter polders are constructed for the purpose to protecting inhabitants from flooding. The total area protected by these polders is about 50 thousand hectares. The period that saw the most intensive used of the polder systems was that from 1970 till 1999.

The Delta area has become a polygon for the development of large state farms for the production of grass powder. Every year, the production of this crop has increased and now stands at around 35-40,000 tons. The product is used for animal farms. Historically, dried grass farming consumed a sizeable amount of energy and fuels:

400 kg of diesel fuel for each ton of grass powder. For Soviet economy, however, this factor did not serve as an impediment, and until 1989 (the last year in which Lithuania constituted a Soviet republic) their annual production of this product was about 39,000 tons.

After the restoration of Lithuania's independence in 1990, the state farms were abolished, and the grass flour industry collapsed. Since then, most of the meadows have been privatized. All households and new farmers *had* been involved in the slowly evolving *market economy.* Cattle herd declined by 3-4 times. Consequently, the need for hay decreased, and part of the grasslands was abandoned. It was in this period that a rapid change in the composition of the grass began. When the meadows were cut three times a year, the grasslands were dominated by dwarf grass swards; when the cutting was reduced to once a year - or not cut at all - the quantities of tall grasses prevailed.

Polder systems have a significant impact on fish stocks. During the spring floods, an abundance of fish entering into the polder channel systems, where they establish the scar areas for spawning. Many of the fish, however, remain in the channels even after the floods. The excess water is pumped from the polder channels using propeller pumps. The use of this type of pump kills or injuries the majority of the fish. In light of this, Archimedes type screw pumps have been installed in the past decade to use instead of propeller pumps.

The rapid changes in human activities have had a negative impact not only on the delta but also on the Curonian Lagoon ecosystem. The meadows of the Nemunas River floodplain, with its specific channel system in the spring, arean important place for fish spawning, but the abandoned grasslands and some inoperative polder systems have created slightly unfavourable conditions not only for fish, but also for birds. A remarkable decrease of migrating water birds has been observedin recent decades, resulting from the changes in the land-use regime that have caused damages to this key avian area.

Floods

The Nemunas delta floodplain extends from Rambynas hill to Kursiu lagoon. The inundated area covers 605 km2, including 402 km2 area in Lithuanian territory (on the right side of the Nemunas River). The right side flood outline extends for up to 5-6 km from the river channel. Meanwhile, the flooded area in the Kaliningrad region territory (on the left side of the river) is less, comprising the 1-1.5 km band between the Levoberoznoje settlement and the town of Rusne, and a band about 5 km wide along Skirvyte, from Rusne to Kursiu lagoon. This entire aforementioned flooded area can increase considerably in the event of levee failures.

The greatest flooding is observed usually in Spring and is the result of melting snow. However, autumn or winter flooding is not unusual, often resulting from both rainfall and snow melt. Additionally, depending upon meteorological conditions in the Curonian Lagoon, the wind-induced surge effects may increase the water level in the coastal area in this period. Flooding usually submerges about 50 km of roads.

Three inundation phases of the floodplain can be distinguished. The first phase is characterized by overspilling of the natural banks. Water overspills low-lying drainage networks when the flooding starts (the Gege, Veizas, Leite tributaries) and further the higher altitudes of the floodplain. Inundated territories are isolated for the first time; water is either stagnant or in movement. As the rise in the water level progresses, the adjacent inundated territories become joined and a concentrated stream flow is formed. This phase starts around mid-March and lasts 2-3 days, during which the water depth on land reaches 1 to 3 meters. The second phase is a continuity of the previous phase and lasts 3-4 weeks. When the water level starts to drop off, the continuous stream flow is disrupted and, as previously, separate floodplain basins appear. The transition period between the second and third phase is generally unclear and its duration is difficult to determine. Approximately 1.2 km3 of water volume can be accumulated in the floodplain. It should be noted that the water flow volume of a 100-year flood on the Nemunas river is about 18 km3, and the capacity of Kursiu lagoon is 6.2 km^3.

The floodplain meadows of the Nemunas delta serve as a trap for sediment deposition, and may significantly decrease sediment transport to the Curonian Lagoon. Several years of monitoring of the sediment transport and the application of a quasi-two-dimensional, depth-averaged hydrodynamic model (Rimkus and Vaikasas, 2010; Vaikasas 2010) showed that an estimated 10,000-90,000 t/year of suspended sediments in the form of clay and silt are deposited in the meadows.

Further investigation of the floodplain soil components have shown that the sedimentation of suspended particles in the Nemunas valley is important for the composition of the surface layer of soil and for the fertilization of the meadows. Such components improve the mechanical and chemical soil makeup and also lessens the amount of fertilizers and organic materials reaching Kurshu Lagoon and the Baltic Sea. (Vaikasas and Dumbrauskas, 2010)

The occurrence of ice on the Nemunas delta river depends on climatic and hydraulic conditions. Ice phenomena there are observed every year. The mean duration of ice cover on the Nemunas river and its branches lasts anywhere from 10 to 97 days. The longest ice cover period may last up to 152 days, while the shortest may involve no days of ice coverage at all. On the rivers' lower stretches, this coverage occurs three days earlier than on the upper delta. The mean ice thickness of the Nemunas river and its branches varies from 25 to 44 cm. The maximum value was recorded in 1969 (83cm). The highest ice thickness is observed in late February and early March. The river's morphological features, such as constrictions, bends, reduction in slope, branches, cause ice jams in the Nemunas delta.

The formation of ice jams in the delta river branches depends generally on the ice cover conditions prevailing in Kursiu lagoon. If the break-up of ice cover in Kurshiu lagoon starts later than in the Nemunas, the ice jams often take place at the mouth. The same phenomenon is observed when floating ice influenced by West or Northwesterly wind on Kursiu lagoon accumulates at the mouth of the Nemunas.

It is clear that ice thickness is not a decisive factor in the formation of ice-jams; rather, the main factor causing such formations is the ratio of drifting ice surface to open water surface. Ice jams do not generate major floods in the delta territory, since the floodwater bypasses them and overspills the riverbanks or low embankments of the right side floodplain. The values of water level increase resulting from the ice-jams varies from 0.4 to 2.5 m. The flood disaster in the Spring of 1958 in the delta and around the Kursiu lagoon coast was the worst in 50 years (Rainys, 1961). It happened when the river was ice-free. The flooding began a month later, as usual.

The water equivalent of snow has varied from 65 mm to 109 mm, and the depth of soil frost has varied from 65-95 cm. The maximum observed discharge was 6580 mc/s (the highest in 100 years).

The Klaipeda-Strait connecting Kursiu lagoon to the Baltic Sea passes at a rate of only 4,500 m3/s. (Gailiusis et. al., 1996). This insufficient capacity raised the water level in Kursiu lagoon by 1.64 m above the Baltic Sea,consequently flooding a large coastal area. Over 100 km of levees were overtopped.. The water depth above their crest was 10-15 cm, and in some cases reached 20-30 cm.

The situation worsened during the flood's recession, when a storm produced a wind-induced surge. In the south-eastern part of the delta, 2 m high waves damaged the levees, resulting in the flooding of considerable portions of land.

Protected areas

As was said before, the Nemunas lowlands and Delta area has several protected areas. The main one is the Nemunas Delta Regional Park, established April 1, 1997, and which makes up part of the Natura 2000 network of natural habitat protection areas and is also a Special Protection Areas for birds. This park occupies about 80% of the delta areas right of the river's coast. Under the

Ramsar Convention criteria, the delta was added to the 'wet areas of international importance' list in 1993.

The Nemunas River Delta may be considered the most important wetland complex in Lithuania and one of only a few close to natural estuaries in the Baltic Region. The seasonally inundated natural meadows and grasslands of the Nemunas floodplain, ox-bow lakes and ponds, raised bogs, fishpond systems and wet forests make this territory one of the of the most unique areas in Lithuania. Thousands of migratory birds can be seen in the lower part of the delta in spring and autumn, especially species of water birds – among them, a number of vulnerable nesting bird species. Rusne Island is a haven for 270 of the 325 Lithuanian bird species. The delta is very popular among birdwatchers. Several times a year they coming here to compete to see who can observe or hear the greatest number of species of birds. The wetlands are also an ornithological paradise. Vente Cape is a small peninsula at the eastern coast of the Curonian Lagoon. This is an ideal geographic location for catching and ringing birds. Up to 300,000 birds per day fly through Vente Cape alone..

The Nemunas delta is famous not only for its natural riches. Here, one can also enjoy its interesting architectural heritage. The Vente and Uostadvaris lighthouses are the area's oldest monuments of technical architecture, as is the Uostadvaris pumping station (Now employed as the Rusne polders museum). (Svažas 2009).

References

- Basalykas, A., (1961), "The lowland of Nemunas delta", in *Geografinis metrastis*, IV t. 5-33 (in Lithanian)
- Bitinas, A., at all. (2002), "Geological development of the Nemunas River Delta and adjacent areas", *West Lithuania Geological Quarterly*, Vol 46, n. 04
- Gailiusis, B., Kovalenkoviene, M., Kriauciuniene, J., (1993), "Hydrological aspects of development of Klaipeda harbor", *Journal of Lithuanian academy of sciences*, n. 3, 73-78 (in Lithanian)
- Jablonskis, J., (1994), "Runoff of the Nemunas during the 180 years", *Journal of Lithuanian academy of sciences*, n. 4, 19-32 (in Lithanian)
- Jablonskis, J., Jurgeleviciene, I., Juskiene, A., (1993), *Hydrography of the River Nemunas*. Vilnius (in Lithanian)
- Lasinskas, M., Burneikis, J., (1961), *Runoff of river Nemunas*, Kaunas (in Russian)
- Macevicius, J., (1972), "Distribution of discharges in Nemunas delta", in *Hidrometeorologiniai straipsniai*, t. 5, 25-33 (in Lithanian)
- Macevicius, J., (1993), "Water levels in river Nemunas delta", *Journal of Lithuanian academy of sciences*, n.3, 11-22 (in Lithanian)
- Rainys, A., (1961), "Extreme flood of 1958 in Nemunas delta area and Kursiu lagoon", in *Geografinis metrastis*, IV t. 163-175 (in Lithanian)
- Rainys, A., (1973), *Dinamic of floods in Nemunas delta*, manuscript of report, Vilnius (in Lithanian)
- Rainys, A., (1991), "Runoff of main branches of Nemunas delta (Gilija and Rusne)", in *Regionine hidrometeorologija*, t. 14, 12-16 (in Lithanian)
- Rimkus, A., and Vaikasas, S., (2010), "Possible ways to improve sediment deposition in the Nemunas Delta" *Hydrology Research*,Vol. 41, n. 3-4, p. 346-354
- Svažas, S., et al. (2009), *Nemuno deltos regioninis parkas*, Vilnius: Akstis, 2009. – 72 p. (in Lithanian)
- Vaikasas, S., (1993), *Development and application of hydraulic modeling for investigation of polder systems*, Ph.D. thesis. Institute of water management, Kedainiai, Lithuanian (in Lithanian)
- Vaikasas, S., (2010), "Mathematical modelling of sediment dynamics and their deposition in Lithuanian rivers and their deltas (case studies)", *Journal of Environmental Engineering and Landscape Management* Vol. 18, n. 3, p. 207 – 216
- Vaikasas, S., and Dumbrauskas, A., (2010), "Self-purification process and retention of nitrogen in floodplain of river Nemunas", *Hydrology Research*, Vol. 41, n. 3-4, p. 338-345

PO

Maria Chiara Tosi

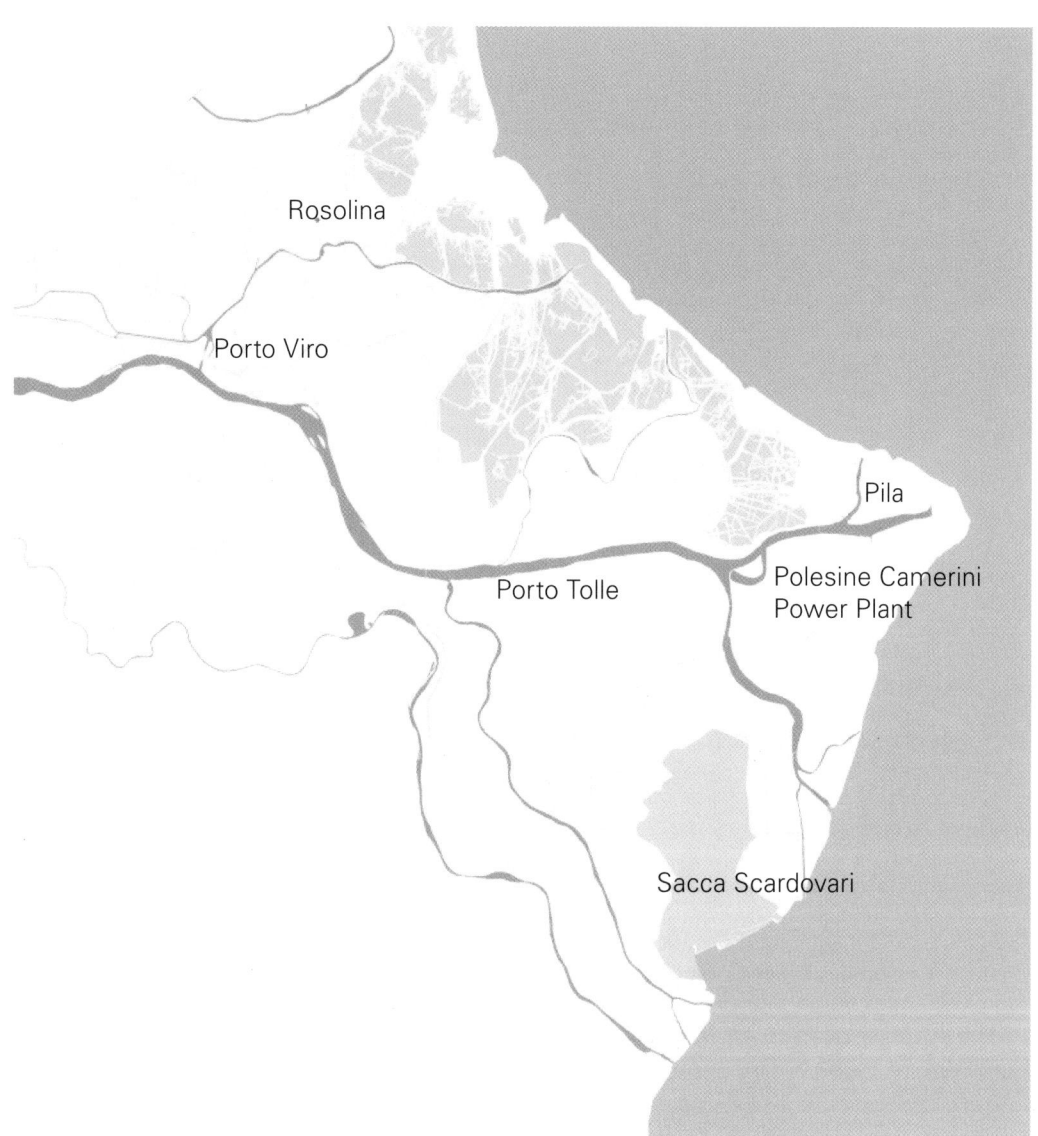

In a constant search for balance

As with many other deltaic territories, in the Po Delta land and water, man and nature are opposed and integrated, conflict and assimilate each other in a process as old as the delta's own formation. (Bertoncin 2004)

Instability, incessant change and a constant search for balance are the normal conditions of this territory, which only at first glance may appear to have been unchanged for centuries (Bondesan 1990).

On the contrary, it is precisely the uncertainty of its geographies, be they physical, social, economic or cultural, that most characterize this area.

Conceptions of the area as a land of conquest and a hopelessly depressed area have long dominated the collective imagination of the Po Delta, today threatened by problems that jeopardize its balance once more. Seawater intrusion, coastal erosion, scarcity of fresh water, increasingly extended periods of drought, marine eustatism exacerbated by the pressures of climate change (Ministry of the Environment 1995, IPCC 2012), are pushing for a reshaping of the geography of this area, found once more to be fragile but at the same time harboring a great potential for innovation.

As the end of the Po river, the delta suffers from the constraints of the catchment area upstream, and of the many civic, agricultural and industrial activities developed along the 650 kilometres of a strategic part of a land inhabited by nearly 15.7 million people and where 46% of the jobs in Italy are concentrated. This area is the controversial result of a complex flow of intentions, policies and projects, and of the natural transformations that these imitate; it is the result of an enormous human effort that over the centuries has descended upon this great river basin, reclaiming, clearing, moving the course of the river and the soil, building small and large dams in the mountains and digging gravel from the river bed in the plains, urbanizing and channeling water for domestic, industrial and agricultural uses, shaping the territory and allowing the delta to have been formed into what we see today. (Day et al. 2005)

One of the largest wetlands in Italy, declared a UNESCO World Heritage Site in 1999, it covers an area of over 786 km2, two thirds of which consist of districts and drainage basins and one third consisting of lagoon or valley environments.

Much of this land is under sea level (2-3 meters), and its drainage is guaranteed by thirty-six water pumps which every year pump one billion cubic meters of water at an average annual electrical expense of €1,200,000: these systems, along with some 600 miles of canals, 480 km of river embankments, 80 km of first-tier sea defence embankments and dozens of kilometres of second-tier sea defence embankments, ensure its safety. (Colombo, Tosini 2009)

The Po Delta is distinguished from its surroundings by the fact that it juts out some 12 km from the coastline into the Adriatic Sea, and also by its small population (73,000 inhabitants) and low settlement density (93 inhabitants/km^2) clustered around a few centres (Porto Tolle, Porto Viro, Rosolina, Loreo), small towns (Gorino Sullam, Boccasette, Pila, Scardovari, Santa Giulia) and many scattered houses built along the roads during the period of agrarian reform.

It is also marked by the strong uniformity of its agricultural landscape composed almost entirely of large expanses of property used for a limited number of crops (corn, soybeans, rice, corn grain and silage, beet and alfalfa), broken only by the subtle and seemingly endless lines of the delta's banks, which together with the coastal dunes are the only high points from which to view the territory's vast horizon.

The long coastline, where land and sea merge in this important lagoon and valley landscape that extends over 16,000 hectares, 10,000 of which are occupied by lagoons and 6,000 by solid land, is a stopping place for countless species of animals.

It is increasingly the destination for an attentive tourism, with great potential for expansion; however, the coastline is also disputed by hunters and fishermen and is a major source of income for the local populations.

Water

The never-ending search for a balance between water and land in the Po Delta has led to the development of techniques to ensure the survival both of the people and of the various economic activities: the exploration of safety devices such as defence embankments to protect against the flooding of the various branches of the Po and against sea storms, and different forms of drainage to keep the land dry and workable represent areas that for centuries have continued to receive much work and attention. This process of gradual reduction of the delta's hydraulic insecurity has tended to bend its environmental conditions through a growing artificiality and technological control of the territory, a progressive reduction of those wetlands susceptible to the influence of the tides and floods, a gradual but inexorable environmental hardening and impoverishment of the territory, which increasingly exists as a mere support to the chosen economic practices and activities: all this leads us to affirm that the Po Delta is trying to withstand climate change by exploring only partially a possible adaptation.

The construction of the modern delta, the complete re-designing of its geography, began with the diversion of the Po river towards Porto Viro (the Porto Viro cut, ca. 1598-1602). From that time and for roughly the next hundred years, beyond the dune-swept ribbon of sandy nature hardly recognizable today, near which winds the route of State Road Romea n. 309, the alternating phenomena of floods, overflowing and sedimentation have led to the accumulation of lands that now make up the delta.

An attempt was made to counter these events of the Po's flooding, compounded by high tides on the Adriatic, by multiplying the river embankments and sea defences, using the most technologically advanced water lifting devices.

Today, the rising sea level and the low river discharge, in the lean periods unable to ensure Pontelagoscuro more than 330 m/s - which the Po Basin Authority judges the necessary minimum to ensure the Vital Minimum Downstream outflow (DMV) - are the main causes of the intrusion of salty sea water that over the past decade has repeatedly come as far as 20 km up the Po river from its mouth, increasing the water's salinity, and consequently that of the land, destroying the crops (especially rice), and making it considerably difficult to ensure access to drinking water for over 15,000 people. Along with this, other exogenous phenomena such as a reduced sediment contribution in the river (Dal Cin 1983), caused by forestry works, the excavation of gravel and sand from the river bed and the construction of mountain basins, and also coastal erosion, make this territory even more fragile.

Responses to these elements of fragility have once again been sought in technology, including such solutions as the construction of anti-salt barriers along the main branch of the Po and of storage basins for fresh water. A response that reveals a slight shift towards progressively adaptive interventions, more willing to communicate with the environment and the landscape than the interventions of even recent history (Colombo, Tosini 2009), and capable of reversing the general progressive reduction of delta's wetland surface areas, whether valleys or lagoons. The latter, despite their immediate reduction, have continued to carry out a fundamental supporting role to the economic activities, which involve a significant number of companies and employees and a high volume of product sold. The clam fishing industry has been particularly involved: beginning here in the mid-80s, its production increased strongly over the following decade, to be later reduced partly due to thermal stress caused by the Polesine Camerini thermal power plant, and partly as a result of the lengthening of the low water periods, and finally, following the increase in pollutants that enhance eutrophication (Simeoni Corbau 2009; Marinov et al 2005).

Lagoon fish breeding, too, is quite common in the twenty-two private valleys covering roughly 8,600 hectares, and this despite the considerable difficulty involved in running fish farms, since subsidence causes water levels to remain constantly below sea level, on account of which lagoon businesses are obliged to use draining systems that drain excess water, culverts and siphons to divert fresh and salt water and water lifting plants for fresh water. Characterized by few employees but high production, lagoon fish breeding is flanked by the most significant wildlife and hunting activities. In the valleys, in fact, about 5,300 acres are allotted to hunting, practiced there by more than 2,000 hunters, who bitterly opposed the creation of the Po Delta park.

Land

The continuous process of taming the water and of reducing it to a technological space has allowed new territories to be built: land to cultivate, but also land to own and exploit by those who do not live in the delta. The State, allied to strong actors such as large landowners living outside the delta, has considered the land a strategic resource for solving economic and social problems, besides political ones. But the processes of territorial transformation have not always had positive results.

A first process involved the launch of agrarian reform (ca. 1951-1960) and the reclamation of a considerable amount of hitherto swampy territory (60,000 hectares), the natural place of mediation between the land and the sea. Though it created approximately 6,000 new landowners (Milan, Perini, Tognon, 2004) and greatly improved the situation of extreme hardship of the farm labourers that characterized the area, it nevertheless failed to develop a new rural society, leaving instead one that continued to remain tightly within the grip of large private capital. (Sereni, 1966)

At the same time, this territory has been marked by the replacement of valleys and of groves of reeds, that complex system of marshy environments typical of deltas with dry and cultivatable lands, forming a densely-knit land of farm plots, a support for the numerous houses scattered among the many farms and a formerly unknown network of roads. The vast and sweeping new landscapes brought about an impoverishment in the environmental quality and in the variety of the production activities: in those years there disappeared all the job types arising from the use of products derived from wetlands. A process of simplification that has continued down until today, with significant losses in terms of biodiversity: from the early 70s until today about 19,000 hectares have been transformed from complex cultivation and land parcelling systems into agricultural areas for intensive farming.

A second process involved the extraction of methane gas and methane-water, and a consequent lowering of the soil level.

The extraction of methane gas and water from about thirty wells in the delta began in 1938 and continued for the next two decades, until it was suspended by the government in 1961. However, since 1951 the mining-induced subsidence raised an outcry over the constant flooding that ensued: the wild drilling into the soil, besides jeopardizing many crops as a result of the entrance of the gas-separated waters into the irrigation and drainage networks, caused significant subsidence phenomena (Borgia, Brighenti, Vitali, 1984), putting in crisis the entire reclamation network, which found it gradually more difficult to overcome increasingly irregular ground levels, and to deal with increasingly salty water.

Though subject to these fragile conditions, the local economy, tightly linked to the amount of available land, painstakingly managed to chart its own path of development, which today increasingly aims to re-centre production on the territory's own prerogatives.

The agricultural sector, historically characterized by the prevalence of large companies oriented towards increasing production and supplying raw materials for the food industry, has more and more seen smaller companies interested in experiencing the results of agricultural research particularly in the horticultural and floral sector, and in processing products and promoting the characteristic, quality products of the region. It is no coincidence that since the 90s rice growing has become increasingly important: there are currently more than 9,000 hectares devoted to this crop, which in 2009 achieved PGI status (Protected Geographical Indication).

With regard to tourism, the delta's characteristic variety of landscapes is a point of excellence along the undifferentiated band of the Italian Adriatic coast, largely dedicated to mass tourism. This is an opportunity for the Po Delta to formulate a more complex touristic offering: from traditional beaches such as Rosolina, or elite ones such as Albarella, to innovative forms of ecotourism, agritourism, and fishing tourism able to enhance the environmental and landscape potentials of the delta. Naturally, it is a small sector, though one with great possibilities for improvement and development.

Even the feeble efforts at reform that can be detected in the variety of productive activities testifies to the attempt, before an intrinsic territorial fragility, to create a positive perspective where integrating the territory's specific qualities is the keystone of local development.

Who governs the Po Delta?

The conflicts and disputes, and the close interrelationship between water and land that have characterized the creation of the Po Delta, are well represented in the events surrounding the Park's creation. Over the last quarter century, the Park has seen a succession of very different ideas, polarized around two major issues: protecting the delta's fragile environment and maintaining and increasing its economic activities.

From the first notions of an interregional park in 1972, until the approval of the Delta Area Plan by the Veneto Region in 1994 and the establishment of the Po Delta Park of Veneto in 1997, proposals and counterproposals to protect all or parts of the park followed one after another, triggering the opposition of most of the players working in the delta and creating a deep rift between the territory lived by the local population and the territory designed by the regional governing body. This partly explains why the area protected by the Park is a very small portion of the territory, corresponding to the watery areas of the valleys, lagoons and branches of the Po, roughly totalling only 12,000 hectares, and why the Environmental Plan of the Park meant to define the various forms of protection and intervention, after a succession of decisions to narrow - and then widen - the protected nature areas, after the submission of a Preliminary Document (2008), Strategic Environmental Assessment (2009) and draft Plan (2010), has yet to be approved. This situation was partly ameliorated by the 1992 Habitats Directive which identified many parts of the delta as belonging to the SIC (Sites of Community Importance) to be included within the Natura 2000 network. The drafting of a management plan has been underway for these areas since 2010.

The regulatory gap left in the territory by the non-approval of the Parks' Environmental Plan, is flanked by the weakness of the other numerous plans and projects developed for the Po Delta by the local governing institutions. This weakness is due mainly to their lack of integration, but also to the segmentation and fragmentation that besides slowing down their effectiveness, has opened the field to all those external entities who intend to transform the territory, considering it a mere resource to exploit or use, a territory seen as backward, depressed, marginal, peripheral, lagging behind in development and with little endogenous capacity for emancipation.

The events related to the exploitation of the delta's energy potential are an important part of this story. In the 60s, when methane gas extraction was suspended, discussion began about creating the largest thermal power plant in Europe. Once an agreement was reached between ENEL (the national entity that manages electricity in Italy) and the municipality of Porto Tolle - the largest municipality of the Delta - construction began in 1973 on the Polesine Camerini thermal power station that became active in 1984, supplying about 8% of the Italian energy demand. The reduction in the number of workers and the first signs of the impacts on the environment (from the heating of the water with a significant impact on the cultivation of clams, to the gas emissions with consequences on the citizens' health) led environmental groups to strongly oppose the Central power plant, obtaining a sentence against ENEL in 2006 for pollution and the closure of the plant in 2010. Today, it is slated to be converted to processing clean coal, but production startup has not yet been greenlighted by a positive environmental assessment.

Another significant story involves the LNG rigasification plant at Porto Levante, an offshore platform 17 kilometres from the coast near Porto Levante which can regasify up to a maximum of 8 billion cubic meters per year. In the hands of owners outside the delta (Qatar Petroleum, ExxonMobil, Edison), the LNG terminal started operating in 2008, creating hundreds of new jobs and awarding a 12 million euro compensation to the surrounding territories. The consequent cooling of the waters and air pollution reported by environmental groups have led to some investigations by the local public prosecutor's office.

In the wake of these events came the 2008 declaration - immediately withdrawn - by the Veneto region President, identifying the Po delta area as a suitable location for a nuclear power plant.

This leads us to conclude that in the Po Delta, the interplay of forces and conflicts between outsiders and local actors continues to dominate the scene: on one side are the local actors whose innovative proposals make protecting and adapting to the environment one of their objectives, seeking to convey a new image of the territory; on the other side, ENEL, AGIP, the large agricultural producers in the valley and the landowners, all of whom continue to act with a view to outsourcing the territory's benefits, regardless of the constraints imposed by the mainly hydraulic- and environment-related vulnerabilities. If there is a different perspective possible today for the Po Delta, it seems to lie in greater coordination, communication and in the sharing of objectives and results among the local bodies, and in the ability to adapt to a fragile environment subjected to major changes. Dismantling the region's image as a forgotten periphery - which has dominated the mentality of those who believed themselves capable of governing it - by pushing for a closer integration between types of expertise and between local and external forces, could constitute one of the possible great innovations for this area. An innovation that must also be sustained by thinking differently about the delta's maintenance: it should not continue to be cared for only by the set of technical apparatuses, but also by society, in its various forms of local action.

References

- Bertoncin, M., (2004), *Logiche di terre e acque. Le geografie incerte del delta del Po*, Cierre, Padua
- Borgia G. C., Brighenti G., Vitali D., (1984), «Estrazioni di fluidi dal sottosuolo e subsidenza del delta del Po», in *Atti della III Tavola Rotonda, Sezione Geologica*, 1982, Tipografia Compositori, Bologna
- Bondesan, M., (2009), (ed), *L'ambiente come risorsa. Il parco del Delta del Po*, Spazio Libri, Ferrara
- Colombo, P., Tosini, L., (2009), *60 anni di bonifica nel delta del Po. 1950-2010*, Papergraf, Padua.
Dal Cin, R., 1983. "I litorali del delta del Po e alle foci dell'Adige e del Brenta: caratteri tessiturali e dispersione dei sedimenti, cause dell'arretramento e previsioni sull'evoluzione futura", in *Bollettino Società Geologica Italiana* 102, 9–56.
- Day, Jr J. W., Abrami, G., Rybczyk, J., Mitsch, W., (2005), "Venice Lagoon and the Po Delta: system functioning as a basis for sustainable management", in Fletcher, C.A., Spencer, T., *Flooding and environmental challenges for Venice and its lagoon: state of knowledge*. Cambridge University Press, Cambridge
- IPCC, (2012), *Managing the Risks of Extreme Events and Disasters to Advance Climate Change Adaptation. Summary for Policymakers*.
- Marinov, D., Zaldívar, J.M., Norro, A., Giordani, G., Viaroli, P. (2005), *Integrated modelling in coastal lagoons part b: lagoon model Sacca di Goro case study* (Italian Adriatic Sea shoreline), In: European Commission DG Joint Research (Ed.), Centre Institute for Environment and Sustainability Inland and Marine Waters Unit
- Milan, D., Perini L., Tognon, C. (2004), *Progettare la terra, progettare la società. L'attività dell'Ente Delta Padano negli anni 50*, Biblioteca del parco, Rovigo
°- Ministero dell'Ambiente, 1995. *First Italian national communication to the framework convention on climate change. Vulnerability assessment and adaption measures*.
- Sereni, E. (1966), *Capitalismo e mercato nazionale in Italia*, Editori Riuniti, Rome
- Simeoni, U., Corbau, C., (2009), "A review of the Delta Po evolution (Italy) related to climatic changes and human impacts", in *Geomorphology* 107

RHINE

Taneha Bacchin

Introduction

The territorial system of the Rhine-Meuse-Scheldt Delta, located between the Netherlands and Belgium, is characterised by the omnipresence of water in different environments. From the North Sea coast to the delta estuaries, this territorial system is formed by the confluence of three major rivers from northwest Europe: the **Rhine**, **the Meuse and the Scheldt**. The larger rivers of the Rhine and the Meuse are divided into Nederrijn-Lek, Waal-Merwede and IJssel, where river clay deposits are dominant. In the northwest region, the delta includes the former Zuiderzee and part of the marine clay deposits in Friesland, and in the southwest, it receives the contribution of the river Scheldt. In the northernmost part, the influence of the North Sea is predominant, where the delta mainly contains marine clay deposits (Nienhuis, 2008). The delta measures 41,543 km², of which large parts of its surface lie below sea level: the lowest elevation is -6.74 m at Nieuwerkerk aan de IJssel in the west of the Netherlands. The lower plain is protected by coastal dunes and an extensive system of dams and dikes. Safety levels are 1/10,000 years for the coastal defence and 1/1,250 years for river dikes. The design discharge is 16,000 m³/s for the Rhine River at Lobith, and 3,800 m³/s for the Meuse River at Eijsden (Schielen and Havinga, 2010); the annual rainfall is 800 mm.

Physical and socio-economic features of the delta

The dynamics of land and water have produced complex landscapes in the three rivers' estuaries, which transition from saltwater to freshwater, from low to high tide, and from dry to wet lands, evolving into a series of rich and diverse ecosystems. A unique environment of brackish marshes, mud flats and canals are characterized by the low-lying landscape and the higher flat land of water and green spaces in the central part of the delta. Moreover, the delta presents a contrasting range of landforms not only between the respective landscapes of Flanders and the Netherlands, but also between polders and sandy soil areas.

The delta's central region is a densely populated, transnational area. The area includes the Dutch Randstad conurbation (Amsterdam, Rotterdam, The Hague, and Utrecht) in the northern portion with an average population density of approximately 1,500 inhabit/km², and the Flemish Diamond (Antwerp-Ghent) and Bruges conurbation in the southern part with an average of 800 inhabit/km². The metropolitan landscape of cities, towns and industries is concentrated in a ring enclosing a central system of blue-green open spaces, and a peripheral zone nearby, canals and ports line /face the delta. The agglomeration of built-up areas – of small, medium and large size cities – within short distances makes the Rhine-Meuse-Scheldt Delta one of the most highly urbanized regions in the world. In the Netherlands, this feature has been supported by steady population growth for centuries. There was an average increase of 3 million inhabitants in 100 years before the 20th century; afterwards, exponential growth occurred, increasing the population from 5 to 16 million. However, future scenarios indicate growth to 17 million inhabitants by 2035 and a subsequent decrease to 16.9 million by 2050, representing a major break in the current trends in the Netherlands (Nienhuis, 2008).

The economic welfare of the delta region is structured in different activities and economic sectors. The highest percentage of employment comes from services, followed by industry and then, with the lowest contribution, farming and fishing. The sub-regions of West Flanders and Zeeland comprise the largest part of the primary sector, while the tertiary sector is concentrated mostly in South Holland and Zeeland. Ports are located at the edge of the delta, including the two largest gateways in Europe (Rotterdam and Antwerp), making it one of the major seaport hubs in the world. Additionally, historical heritage and the natural environment play an important role in the way the territory is characterized by and used for cultural activities, resources and recreation.

Designing with Nature / Planning with Water: the evolution of the Rhine-Meuse-Scheldt Delta landscape

The interplay between infrastructure and the

dynamics of nature over land and water has guided the territorial development in the past two centuries. A sequence of historical events triggered the design of new concepts and visions, that were followed by specific spatial configurations and governance arrangements geared towards standards for ensuring the delta's future.

After the Zuiderzee Flood of 1916, when dikes collapsed under the stress of a winter storm, the former plans for the reclamation of the Zuiderzee were revised. In 1920 the project began, after the Zuiderzee Act of 1918 established the main goals for the Zuiderzee Works: protecting the central Netherlands from the fluctuation of the North Sea, ensuring the provision of food through new agricultural land, and improving water management through the creation of the Ijsselmeer lake from the previously uncontrolled salt water inlet. In addition to creating Lake Ijsselmeer, the Zuiderzee Works also created 1,650 km2 of new land. The Works were concluded in 1933 with the opening of the Afsluitdijk dike. The next phase involved reclaiming new polders by damming and draining portions of the Ijsselmeer. Before World War II, the Wieringermeer polder was reclaimed from the Zuiderzee, and was the first of the five polders designed to supply new land. From 1931, four villages were formed in the polder: Slootdorp, Middenmeer, Wieringerwerf and Kreileroord. The second polder created was the Noordoostpolder, comprising the former islands of Urk and Schokland.

However, in 1937 studies conducted by the Department of Waterways and Public Works (Rijkswaterstaat) showed that different parts of the Netherlands were under risk of flooding. The study identified difficulties in increasing the protection levels of the original system of dikes near the river mouths.

The period after the World War II was marked by reconstruction efforts and further reclamation of land to be assigned to agriculture and habitation. Seawater intrusion into the groundwater caused a loss of arable fields in the region subsequent to a deepening of waterways near the coast and the subsidence of land following the construction of previous polders. The forth polder of the Zuiderzee Works – Flevopolder – was built in 1950 and created 1,000 km² of new land. The project included large urban settlements to relieve the housing shortage; the cities of Lelystad and Almere belong to this area. The North Sea flood of 1953 was one of the biggest in the history of the Netherlands. Nearly two thousand people died and 150,000 hectares of land were flooded in the provinces of Zeeland and South Holland. A few days after the flood, the Delta commission was established to advise on the execution of the 'Deltaplan' (latter renamed the 'Deltaworks'), created to make the Delta area safe. The construction of the Deltaworks started with the partition of the southwestern delta into multiple compartments via a sequence of dams built in the Zandkreek, Krammer, Grevelingen, and Volkerak rivers. By 1958 the first Deltawork, the storm barrier in the Hollandse Ijssel river, was finished; it protected the densely populated western part of the country (the Randstad) from flooding. By 1961 the Veerse Gat and the Zandkreek river mouths were closed, thus forming a new lake: the Veerse Meer. After the first Deltaworks, a series of projects were further executed through the Deltaworks programme: the Haringvliet sluices and the Brouwers dam, the storm surge barriers of the Eastern Schelde, Oosterschelde (a partially open barrier), Hartelkering and Maeslantbarrier. Given the economic importance of the ports of Rotterdam and Antwerp, the waterways of the De Niewe Waterweg and of the Western Schelde stayed open.

Over the centuries, there has been increasing pressure on the available land area in the Rhine-Meuse-Scheldt Delta caused by economic growth and urbanization patterns. In the river regions, the consequences of the intensive land use are the reallocation of dikes closer to rivers, land occupation on the river beds, and canalization of streams to improve the efficiency of water discharge, among others. In 1993 and 1995, the Netherlands once more was exposed to flooding, due to increase in the river discharge caused by large volumes of snow melt and rainwater from upstream regions.

Designed dikes were not able to withstand the increased water levels. These events exposed the urgency of developing higher safety standards, and new spatial and political concepts for dealing with the vulnerability of people and assets in the river regions.

Policies and planning in the Delta

In 1996, the former Dutch Ministry of Transport, Public Works and Water Management (now part of the Ministry of Infrastructure and the Environment) presented a new Policy Guideline on Major Rivers focusing on an innovative set of strategies for the development of River Regions in the Netherlands. The Policy Guideline on Major Rivers was structured for prevention, spatial planning and environmental quality objectives rather than for evacuation and reconstruction. The basic concepts mainly introduced measures to create more room for the rivers that also could offer the possibility of preserving and/or improving the spatial quality of the region in the center of the country. A number of position documents were introduced by the Government after the first Policy Guideline. In 2000, the Government's position paper 'Dealing Differently with Water – Water Policy for the 21th Century' showed a reduction on safety levels in the delta and the likelihood of higher risk of flooding due to climate change and land subsidence. In the same year, the position paper 'Room for Rivers' provided the framework for the development of new spatial planning, and technical and governance concepts for preventing flooding in the event of increased discharge levels without further dike reinforcement work. In 2001 the 'Preliminary Agreement on Water Policy in the 21th Century' was published.

International policy visions for the delta are embedded in the policy guidelines developed in the Netherlands. In particular, the partnership between countries along the Rhine, Meuse and Scheldt Rivers allowed the implementation of the Flood Action Plan for the rivers involved after 1998. The European Framework Directive on Water (FDW), applied since 2000, is aimed at enhancing the protection and restoration of water bodies by 2015, and the coordination of efforts between EU Member States. Especially for the Netherlands, situated in the downstream Delta, the international collaboration between countries responsible for upstream and central catchment areas is critical. Disaster management is addressed in the Government's position paper 'Flood Disaster Management Strategy' published in 2003, introducing options for enhancing international coordination, emergency overflow areas, higher flood safety standards, and organizational measures.

Actors

Three key actors in the Netherlands were responsible for the territorial development of the Delta in recent centuries: these are the Directorate of Public Works and Water Management (Rijkswaterstaat), the Delta Programme Commissioner (Deltacommissaris) and the Dutch Water Boards (Waterschappen).

The Directorate of Public Works and Water Management (Rijkswaterstaat) was established at the end of the 18th century to manage the dynamics of land and water and improve safety to a large extent via infrastructural works. In the course of more than two centuries, Rijkswaterstaat has executed railways, navigable waterways, reduced and transformed the coastline and reclaimed over 200,000 hectares of land. Rijkswaterstaat manages the national road and waterways networks, and the water system, including the Dutch part of the North Sea.

The Delta Programme Commissioner was established in the course of the 20th century for the development and execution of the Delta Plans. The Delta Programme is responsible for investments, planning and execution of the Delta Works seeking the protection of population and assets, and prevention from the negative impacts of drivers of future change in the delta. In 2007 the Delta Committee made recommendations regarding the way in which, in the century ahead, water safety and freshwater supply must be managed, taking into account climate change and social developments. The third Delta Programme (DP2013) was presented in 2012, and the now Delta Act went into force constituting the legal basis for the Delta Fund to finance the Delta Works of the future. The Act provides for the appointment of a Delta Commissioner, responsible for ensuring that a Delta Programme is drawn up and implemented and its progress supervised every year. The third Delta Programme highlights the development of multi-layered safety strategies. The first layer of safety is prevention through dikes, dunes, barriers and dams. The second layer is provided by spatial planning to limit the vulnerability and consequences of flooding. To close the safety cycle, emergency management constitutes the third layer.

The framework legislation Water Act focuses on integrated water management practices based on the relationship within different systems; e.g. the quality and quantity of water, surface and groundwater, land use and water uses. The integrated water system approach also includes policies in the fields of nature, environment and spatial planning. In the Water Act, the authorities responsible for the water management in the Netherlands are the State as the authority for the main waterways and the Water Boards as the authorities for the regional waterways, water quality and sewage treatment in their respective regions. Water Boards are among the oldest forms of democratic government in the Netherlands. Hoogheemraadschap van Rijnland is the oldest water authority that is still in function, initially being founded in the 13th century.

Assets and fragilities of the Delta landscape

The natural forces of the Rhine-Meuse-Scheldt Delta were intensely managed by man-made systems for centuries, leading to a complex landscape shaped by spatial patterns of wetlands, embankments, polders, cities, and infrastructures including the two largest ports in Europe. The fragility and strengths of this territory therefore rely on an ability to work with natural processes and socio-economic development over the years. Two thirds of the Netherlands is vulnerable to flooding, presenting high levels of potential damages. In addition, sea level rise will increase the existing problem of salt water seepage in the delta. To overcome the damages of salinization for agriculture, brackish cultivation using salt-tolerant crops provides innovation for arable land below sea level in the Netherlands.

The management of the natural landscape over the course of centuries has caused the loss of biodiversity in estuarine and coastal ecosystems due to pollution and reduced hydrodynamics. The Dutch government via 'The Nature Conservation Act' provides for the protection of natural areas under '**The** National Ecological Network' and the 'European Network of Protected Sites - Natura 2000'. Coastal erosion is being controlled with a novel strategy of extensive sand nourishments. **The Dutch Programme 'Building with Nature' is an innovative research agenda aimed at developing new design concepts for a sustainable management of coastal, delta and riverine regions. Natural processes are integrated into planning and design, enhancing the balance between natural ecosystems and human intervention.** (van Dalfsen, Aarninkho, 2009). As an example, the pilot project under development in the southwestern delta region (Western Scheldt area) studies how the natural flow of sand along the Dutch coast can be used to improve coastal defence systems.

Conclusion

The Rhine-Meuse-Scheldt Delta is a comprehensively unified territory, shaped by the presence of water in natural and manmade landscapes from the North Sea to the delta estuaries. The physical, historical and cultural identities and heritage of cities such as Dordrecht, Rotterdam, Flushing, Antwerp, Bruges and Ghent are characterized by the watercourses as part of the identity image of their city centres.

In the history of the delta the main planning issue was the balance between regional economic development, social well-being and the restoration of ecological values. New policy visions and planning instruments, including 'Room for Rivers', 'Dynamic Estuaries' and 'Building with Nature', aim to sustain regional economic growth by integrating the **dynamics of nature** on coastal and riverine flood defence systems and deliver solutions for recreation demands and urban development. Historical coastal defence plans have caused major environmental impacts on the Dutch and Belgium coastal systems, where the urgency of restoring ecosystem functions is clear. Future proposals for the delta include the reclamation and extension of the Port of Rotterdam and the deepening of the waterways access to the Port of Antwerp in the Western Scheldt. Long term pressures in the delta include sea level rise and higher river discharge levels due to climate change, additional land subsidence, salinization and freshwater shortage. Furthermore, related socio-economic trends such as the increased competition between markets (particularly in terms of port economy), and the shrinking and ageing of the population, among other things, might also strongly influence future scenarios of the delta landscape.

References

- Deltanet Project. Network of European Delta Regions. *Rhine, Meuse and Scheldt Delta*.
- Geuze, A., and Feddes, F., (2005), *Polders! Gedicht Nederland*. NAi Publishers. Rotterdam, The Netherlands.
- Hall, P., and Pain, K. ,(2005), *Commuting and the definition of functional urban regions: The Randstad*, Institute of Community Studies. Polynet Action 1.1. Polynet Parterns
- Huisman, P., de Jong, J., Wieriks, K., (2000), *Transboundary Cooperation in shared River Basins: Experiences from the Rhine, Meuse and North Sea*. Water Policy 2(1-2): 83-97.
- Lintsen, H.W., (1998), *Decline and rise of Rijkswaterstaat. Two centuries of history*, Den Haag: Ministry of Transport, Public Works and Water Management, 24 pp.
- Ministry of Infrastructure and the Environment, Ministry of Economic Affairs, Agriculture and Innovation (2011), *Delta Programme 2012 | Working on the delta*. Translation of the 'Deltaprogramma 2012'. The Netherlands, September
- Ministry of Infrastructure and the Environment, Ministry of Economic Affairs, Agriculture and Innovation (2012), *Delta Programme 2013| Working on the delta*. Translation of the 'Deltaprogramma 2013. The Netherlands, September
- Ministry of Infrastructure and Environment, Directorate-General Water and Rijkswaterstaat, Centre for Water Management (2011), *Water management in the Netherlands*. The Netherlands, February
- Ministry of Transport, Public Works and Water Management, the Ministry of Housing, Spatial Planning and the Environment and the Ministry of Agriculture, Nature and Food Quality (2009), *River basin management plans Ems, Meuse, Rhine Delta and Scheldt – a summary*. The Netherlands
- Nienhuis, P.H., (2008), *Environmental History of the Rhine—Meuse Delta*, Springer Science + Business Media, B.V.
- Programme Directorate 'Room for the River' (2012), *Spatial planning key decision Room for the River: Explanatory memorandum*. The Netherlands
- Programme Directorate 'Room for the River' (2012), *Spatial planning key decision Room for the River: Approved decision*. The Netherlands
- Schielen, R. M. J. & Havinga, H. (2010). *Long term claims on the Dutch river area: handling climate change, safety, navigation and nature*. In A. Dittrich, K. Koll, J. Aberle & P. Geisenhainer (Eds.), River Flow 2010 Proceedings of the International Conference on Fluvial Hydraulics , 8-10 September 2010, Braunschweig, Germany (pp. 1429-1436). Karlsruhe: Bundesanstalt fuer Wasserbau
- Stichting Deltawerken Online (2009), *Deltawerken: Water, nature, people, technology*
- van Dalfsen, J. A., Aarninkho, S. G. J., (2009), *Building with Nature: Mega nourishments and ecological landscaping of extraction areas*. In: European Marine Sand and Gravel Group – a wave of opportunities for the marine aggregates industry. EMSAGG Conference, 7-8 May 2009, Rome, Italy

RHONE

Fabio Vanin

Where

The Rhône delta is a vast alluvial plain (145,000 ha) formed by the accumulation of sediments deposited in the collision between the river and the marine elements of the Mediterranean Sea. It shaped like a triangle whose apex is the city of Arles and whose base is the coast from the Gulf of Fos to Aigues-Mortes. The Rhône River, 812 kilometres long and with a catchment area of 99,000 sq km, is one of the five major rivers in France and the one with the highest discharge (an average at the mouth of 1800 m3/s): born from the Saint-Gotthard Massif, it crosses south-west Switzerland passing through Lake Geneva and then the south of France, through the Alps and the Central Massif, forming a flood valley from Avignon and then widening into the large delta that gives rise to the Camargue region (a Natural Park since 1970).

Downstream of Arles, the river divides into two main branches that pass through the delta apparatus: the Great Rhône to the south-east (length 50 km, 87-92% of the Rhône's capacity) and the Petit Rhône to the south-west (length 60 km, 8-13% of the river's capacity). Following the cut that the two branches of the Rhône make on the territory by flowing into the Mediterranean, the Camargue region can be divided into three parts: the Petite Camargue, a very depressed area west of the Petit Rhône, the Grande Camargue, between the two branches of the Rhône, and the Plan du Bourg, east of the Grand Rhône.

But there is also a second reading of the territory, overlapping the first from north to south. While the northern part of the delta is characterized by agricultural land, particularly devoted to rice cultivation and to the breeding of horses and bulls, the southern part is formed by a special and precious ecosystem made up of marshes and lakes (the largest of which is that of Vaccarès) which becomes, toward the coast, a band of salt works and a cord that over time has gradually hardened in response to environmental pressures.

Due to the river's irregular flow, the siltation of the coast and the increase in surface area (about 1 square kilometre per year), navigating the river has always been difficult and the Rhône delta has no major ports. However, thanks to a series of works, freight traffic along the stretch of river from Lyons to the sea is now equal to roughly 4 million tons per year (mainly ores, agricultural products, chemicals, fuels) and the maintenance and expansion of the network of deltaic waterways, together with the creation of the national nature reserve (1927) and the Park of the Camargue (1970), have allowed the development of many economic activities as well as those of the tourism sector.

The Rhône River and its delta are historically a place of exchange, trade and transport, which over the centuries has favoured the birth of villages, towns, and major cities (including Geneva, Lyon, Avignon), often in strategic positions and with a history closely linked to the river. In the deltaic areas, there are centres smaller than the river's course, the largest of which is Arles.

When

Defined in the Greek-Roman era as "earth in movement", the Rhône delta has always been at once feared and seen as a place of conquest. The first works linked to a real transformation of the territory can be traced back to the construction of the Cistercian abbeys, also called "salt abbeys", which began the first reclamation of the marshes, and to the creation of the Saint Giles Way, the easternmost road of the Carolingian kingdom.

The main link between northern Europe and the Mediterranean, the Rhône River has been one of the main trade routes since prehistory, providing a quick and economical route for goods. However, the physical characteristics of the river and its delta have always made navigation difficult. Until the nineteenth century, flooding, winter frosts and lean summers which could last up to six months a year made navigating the river a very complicated affair. Among the most important works in modern times was the transformation of Aigues-Mortes into a river port (1806) thanks to the construction of the *canal du Rhône a Sètes*, the conclusion of a longstanding attempt to link the river to the sea. The delta also bears the marks of having suffered strong natural forces for hundreds of years, from the sea and from the river. The mobility of certain elements, such as meanders, and the continuous territorial transformations due to heavy flooding, make it a region at once living and fragile. The instability born from these phenomena, together with the combined influence of fresh and salt water led to the construction of defence equipment and water management systems. Thus, from the second half of the nineteenth century, major containment works were carried out to protect against the river's flooding and saltwater intrusion. The reduction of the Rhône, however, also led to a reduction in the inflow of fresh water, which had to be offset by the construction of a vast network of irrigation and drainage systems developed mostly during the nineteenth century and the first half of the twentieth century. This precarious balance is still maintained today thanks to artificial water management, which helps to regulate its level and fight the salinisation of the soils but creates many conflicts.

Over the centuries, man has profoundly changed the landscape of the Rhône delta following a logic of territorial exploitation. The salt industry, rice cultivation and the development of tourism have helped create areas strongly influenced by man, which exist in close contact with natural areas of great value. Starting from the twentieth century, numerous protective constraints were set up around these prized areas, from the creation of the National Reserve of the Camargue in 1927 to the latest European directives.

Who

The Rhône delta is now the subject of many political and economic interests. The bias towards transformations aimed at protecting individuals or different categories such as breeders, farmers, hunters, the tourism sector or that of real estate often come into conflict with those transformations promoted by the Regional Park of the Camargue, the various government departments and by the respective municipalities. The work of rice farmers and salt work managers, for example, has led to the creation of a hydrology of some areas of the delta that opposes the natural trend and the protection of the native species.

Agriculture in the Rhône delta is accompanied by a long history of struggling against the water. Control of drainage and irrigation therefore plays a key role in the soil's development: rice alone, for example, utilizes about 400 million m3 of water per year. Moreover, the introduction of fresh water has affected the water balance and led to historical conflicts between farmers and salt workers. Wheat and shepherding, predominant in the late nineteenth century, were quickly replaced by vineyards and rice crops, which occupy most of the arable land. To this we can add a phenomenon that has intensified today: the increase of large farms (from 100 to 2,000 hectares), due to the low productivity of the soils and the high cost of basic maintenance (irrigation and drainage). The result is an increasing gradient of property size as one goes from north to south.

Salt is another major player in the Rhône delta and has been extracted since ancient times. The so-called salt industry began in the mid-nineteenth century and grew from the 1950s to the 1970s, becoming a large-scale extraction activity. To be effective, however, this required appropriate weather conditions, the use of advanced technologies and production methods, the digital management of the water's movement, and for the harvest to be done by machines. From the urban-territorial perspective, the exploitation of salt helped build real industrial citadels, with a very hierarchical spatial arrangement, as in Salin-de-Giraud, created in 1860 to house the employees of Pechiney and Solvay. Today, the large-scale salt production (about one million tons per year, which is controlled by the *Salins du Midi and Salines - CSME company*) remains an indispensable raw material for the chemical industry and therefore one of strategic importance. Also, since salt is considered a renewable resource, there is almost no limit to its exploitation.

Since the seventies, the entire region of the Rhône delta has been experiencing a strong increase in tourism (currently around one million visitors each year). While at one time the main tourist attraction was the 40 km of coastline untouched by urbanization and synonymous with freedom and wild nature, today the tourist pressure is gradually spreading throughout the delta. To diversify the forms of tourism, local actors have spurred the creation of trails, museums, centres and activities inspired by local traditions and practices such as bullfighting, horseback riding, etc. While these interventions allowed tourists to visit the interior of delta, other activities, such as hunting, have led to a sometimes radical transformation of the territory. The conversion of farms into tourist businesses unrelated to agriculture, and of rice fields and swamps into private hunting reserves, manifests a trend in the direction of a heavily privatized landscape.

However, tourism and environmental protection go hand in hand. Numerous protection statutes covering the Rhône Delta also coincide with important players influencing it. The biological and botanical reserve classified as a National Nature Reserve is managed by the Société nationale de protection de la nature. The Regional Natural Park was instead created in 1970 by private entities and made an institution in 2004. The Special Protection Areas (SPAs) and Special Areas of Conservation (SACs), belonging to the Natura 2000 network, are the result of European directives. The Biosphere Reserve designation was given by UNESCO.

What

The Rhône Delta has always been a fragile and constantly changing territory, because it suffers the twin pressures of the river and the Mediterranean. Today, the large amount of sediment (sand, gravel and silt) that accumulates in some areas, such as the Gulf of Fos, goes hand in hand with a strong coastal erosion, as occurs at the mouth of the Grand Rhône. The rise in sea level due to global warming is threatening the delta plains not only because of the increased flooding but also due to the increased salinisation. To this we must also add a variety of natural phenomena - such as subsidence, heavy storms and the sea surges regularly striking the south-east coast - which make the delta's future even more precarious.

The doubled number of floods to which the delta is subjected is one of the greatest threats to this area. There are different types of floods, ranging from the so-called oceanic floods - from October to March, brought on by westerly winds and characterized by intensive rainfall - to the Mediterranean floods - characterized by their late onset - to combined floods, which have a more sizeable effect. The concatenation of several combined and widespread storms has given rise to the greatest historical flood events, such as those of 1840, 1856, 1993 and 2003.

The coast of the Rhône delta is subject to substantial erosion - an average of 4 meters per year over the last 60 years - caused by structural characteristics (lack of sediment, fine sand and wave motion). In general, erosion leads to a higher risk of retreat of the coastline and flooding, reduces the width of the beaches and constitutes a threat to the dunes, thus threatening the salt industry, as happens in Faraman or the Petite Camargue. In the medium term (about 30 years), the retreat of the coastline in combination with the flooding and the expected increase in sea level could isolate entire areas and major centres.

The tourism and business activities currently operating in the delta, have and in the future may continue to have a direct effect on protective measures, although today such entities are often in conflict with them. The saltpans, the delta's historic activity, are concentrated along the coast, especially in Petit Camargue (10,000 ha) and Salin de Giraud (12,000 ha) and are protected by mixed systems of dikes and breakwaters, sometimes insufficiently so. At Fos-sur-Mer there is an important industrial port to the east of the delta, home to activities related to the oil, iron and steel industries.

The local farms of a certain economic importance are found mainly in the Petite Camargue and in the northern part of the delta, and the dominant production of these farms is rice (64% of the arable land). Tourist activities are concentrated mainly around the seaside, near Saintes-Maries-de-la-Mer, on both sides of the mouth of the Grand Rhône and near the marina of Port Camargue, west of the delta. The ecological richness, the delta's important heritage that is enhanced and protected by the

Regional Nature Reserve and the National Reserve of Camargue and the coastal environment, also generates an appealing "green" tourism that is growing steadily. However, the economic activities described above, which represent a potential and a resource for the delta, tend to create their own defences against environmental threats in an uncoordinated manner.

How

The prospect of sea level rise is one of the main threats to the territory of the Rhône Delta. Protection systems built to protect the delta, between the mouth of the Grand Rhône and the Grande Motte, are occasional and scattered, uncoordinated, built at different times and by different actors. As a result of this finding, the SYMADREM - Syndicat Mixte d'aménagement Interregional des Digues du Delta du Rhône et de la mer - has proposed a new approach based on what has guided the development of the Plan of the Rhône.

The objective of this new approach is to create an interregional strategy of coastal development in the short, medium and long term: a Coastal Plan that takes into account the climate change and aspects of the lands located inland of the coast. This new interregional plan should seek the best, most financially sustainable solutions. As it involves implementing strategies of coastal management, on an interregional basis, the state plays a key role through its decentralized services, and should work closely with the community and the local actors, in the same way it has been done in the Rhône basin. In order to develop coastal protection strategies, SYMADREM also offers a financing plan based on the review of project contracts between the State and the Regions in an inter-regional approach.

The Regional Natural Park of the Camargue also plays a major role in the development of the territory. In addition to protection and monitoring initiatives linked to a territorial observatory that records the region's evolution from an ecological, financial and social viewpoint, the Park Authority is equipped with a team for rural development and land management. The skills and the services provided range from support for farmers who diversify crops, monitor the area, preserve wetlands and resort to practices which respect the environment, to support for associations of agricultural products in enhancing products and skills, to support for the development of tourism related to the use of the natural resource base, to support for the monitoring of programs, plans and strategies for urban and regional planning with a focus on the natural and architectural landscape. In this way, the Park presents itself not only as a subject carrying out an activity of conservation of the territory and the landscape, but also as an active player, a broker with a view to careful development.

References

- AAVV, *Gestion du risque inondation et changement social dans le delta du Rhône - les catastrophes de 1856 et 1993-1994*, Editions Quae, 2006
- Jiménez, J. ; Capobianco, M. ; Suanez, S. ; Piero, R. ; Fraunié, P. ; Stive, M.J.F. *Coastal processes along the Ebro, Po and Rhône deltas*, MEDCOAST 95, October, 24-27, 1995. E. Özhan Ed., pp. 827-840.
- Picon, Bernard, *L'espace et le temps en Camargue*, Actes Sud, 1988.
- Pritchard, Sara B., *Confluence: the Nature of Technology and the Rethinking of the Rhône*, Harvard University Press, 2011.
- Provansal, Mireille, *Rhône Delta-Eurosion Case Study*, IZCM projects and case studies, 2010.

Part Two
Po Delta

Some Iphotesis

The second part of the book offers a series of reflections, investigations and scenarios for the Po River Delta, a fragile territory of paradoxes and dilemmas, characteristics that are shared and experienced by many other deltaic territories. It is for this reason that this work is presented as an interpretative project endeavour which other fragile territories can learn from as a useful reference case.

The primary aim of this work is above all to single out the resources and critical conditions within the Po Delta region and to plan scenarios for their worthwhile utilization.

The actions should contend with the "logic of land and water" belonging to a territory that is both fragile and affected by important climate changes. These characteristics, one inherent to the region under study, the other caused by its transformations over time, are producing significant effects on the places of and ways of living the Delta.

They must determine in which direction the scenarios should be created.

The implicit question is: what is the future for the Po Delta region?

Three assumptions underlie this question:

1. In recent years, both in Europe and at the local level, a strong awareness has arisen concerning the topic of climate change.

2. This awareness has renewed the bond between a territory and the quality of life of its inhabitants. Landscape increasingly represents a factor of primary importance in determining the personal and collective well-being and welfare of a population and, as such, is an economic resource worth investing from the viewpoint of sustainable development. It is a potential value that can be expressed solely by activating and circulating the territory's great quantity of endogenous resources, with the conviction that the will for social, cultural and technological development is closely connected to the will for regional development.

3. These significant factors offer an imperative view for significant changes in project – intervention strategies. It's then important to reflect on such changes while also involving and communicating with the local society.

The starting point of all our discussions is the assumption that coastal erosion, the rise in average sea levels and the salinization of fresh water constitute the three main factors that should inevitably be considered by anyone dealing with the transformations of this area. These kind of vulnerabilities are common to the main deltaic territories in Europe.

Other notable factors characterize these territories as well, such as the ongoing abandonment of these lands by the local populations, the sharp increase of tourism and its impact, and even the transformation and strengthening of the agricultural sector in these territories.

In order to carry the examination of such phenomena beyond a purely technical discussion and re-interpret them, not as threats and elements of fragility, but as opportunities to be adopted along alternative paths of development, it is necessary to change radically the way we envision the future of the Po River Delta.

Starting from such interpretative work, two hypotheses served as the basis of the exploration that was conducted.

The first hypothesis was that the Po River Delta cannot be considered an immobile territory, or a fixed object, but rather an animated and dynamic phenomenon demonstrating an ensemble of composite and continuously changing materials. However, this clashes with one relevant issue: in this territory, it is difficult to experience any real change unless it occurs under very extreme conditions, that is, when the immense power of the water causes tragic devastation. Over time, this unique characteristic has resulted in a collective vision in which instability, insecurity, and depression comprise the main elements through which the territory and its population describe themselves and tell their story, along with an insistent and relentless impulse towards conditions of strength, security and prosperity.

This first consideration led us to re-trace the reasons for this negative image and to investigate certain assumptions, on which the actions of local society are based, concerning its inhabitants: the young and old, entrepreneurs, politicians and administrators, as well as those who work and live their very existence in this territory. An open discussion with the inhabitants allowed us to develop an important tool to better understand and assimilate local knowledge, often difficult to obtain.

The second hypothesis recognizes the Po Delta as a territory where, over a long period of time and through a series of stratifications, some distinctions were necessary for the following different elements: water/land, nature/culture, river/settlements. These distinctions often generated forms of separation among the people involved, and among their specific organisation strategies, economies, paths of development, and the systems of settlement and infrastructure adopted in the territory. Resolving the tension between such opposites has been difficult; today, this tension is one of the main factors generating those conflicting efforts which this territory is attempting to deals with in devising shared strategies for specific paths of development leading to a common goal.

Over time, the different approaches to land and water, and the various strategies and systems of the actors who have shaped this region, have strongly contributed fragmentating both the area's physical form and its organisation. In recent years, some strategies and programs have provided only specific and contingent responses, incapable of granting unified perspectives.

These kind of approaches have produced a sort of territorial zoning which juxtaposes and distinguishes among: agricultural areas, aquaculture plots, marshes and lagoons, built up areas, industrial zones, residential areas and places for tourist infrastructure. In this perspective, each area has a strong specificity and functional autonomy and is thus supported by distinct strategies and group of subjects. This then creates a collage-like array where individual allotments and individual actors struggle to collaborate with one another.

It is from this point that we started our research, intuiting that the untapped potential of this region could be found through an integration of its key functions, for example: combining agriculture with tourism and the production of fresh water reserves for irrigation, envisioning a newly functioning water management system that is more resilient and responsive to ongoing climate changes. Or combining tourism and aquaculture with the goal of fighting salinisation and erosion of the coastline. Such endeavours represent an important conceptual shift: from separation towards integration, from functional specialization to the multi-functionality of the regional territory.

It's through this shift that we see the possibility to trigger the development of economies that are more varied and articulated, more resistant and durable. Approaches to environmental issues cannot and must not overlook this conceptual shift.

A different organization of the territory's functions in conjunction with different landscape designs will be able to deal with these issues and utilize technical constraints as opportunities rather than experiencing them as limitations. Local communities can take on the responsibility for such demands by adopting different and more complex paths of development.

It is only under such circumstances that the prospect of living the Park of the Po Delta as a place of permanent or temporary residence for work, transit, spending free time, or receiving an education, can represent an objective goal rather than a fate to be suffered.

One final issue.

The continual and recurrent process of change that the contemporary territory is undergoing is increasingly appearing to be an intrinsic feature, a specific connotation of a society that counts the notion of "crisis" amongst its paradigms.

Very often change has been characterized as a manifest alteration of the conditions of a territory, as a radical transformation. However, what is new and different can also manifest itself through a progression of minute changes resulting in giant leaps, when the system resists a slow crescendo of efforts until it reaches the breaking point.

This different notion of transformation suggests considering different "geographies of change": whilst, on one hand, it is vital to understand what creates a rupture with the past, and with the spaces and practices that reveal the fragility of tradition and radically transform it, on the other hand, it is also necessary to focus on the everyday process of change that is as minute as it is continuous, pervasive and incremental, transforming the character, role and meaning of the spaces of the territory.

In other words, our attention needs to be steered towards that "articulated, multidimensional and dispersed" process, whose intensity is difficult to measure but that concerns a multitude of forms and modalities, intersecting in different ways with the specificity of places and the constellation of local actors.

We need to focus on what we perceive as "innovative", but also on what we consider to be "adaptation", expanding our notion of 'new' to include what is often dismissed as mere accident, alteration, accumulation and variation.

Amongst the changes that contribute particularly to reconfiguring the space of everyday life, those concerning the climate – including the predictions of future climate change – have an impact not only on the material aspect of territory, modifying it strongly, but also on the ability of a society to imagine its future, a future that appears increasingly remote, uncertain and hard to decode.

Failure to imagine a future in relation to climate change often derives from the assumption that everything will change suddenly at some point, a point which is as uncertain as it is hard to position on a timeline. Such a view fails to consider that our present is already completely immersed in this process of change and that small everyday transformations have already shaped the territories we inhabit.

Constructing scenarios that reveal small and large transformations, and exploring their material consequences, is a way to reduce uncertainty about the future and helps society make informed decisions around the potential opportunities that strategies for intervention might present.

In such circumstances, a specialised approach to understanding what is happening in our territory always proves ineffective. Instead, it is necessary to experiment with "thick" descriptions and ways of investigating that resist attributing objects, spaces and behaviours with exogenous meanings based on taxonomy. Such "thick" modes of investigating acknowledge the multiple roots of the changes in progress, and try to consider simultaneously the economic subjects, the interactions at a micro territorial level, the spaces of local economic activities, and the environmental characteristics that influence the organizational models of both the local economy and the local settlements.

This approach leads us to consider the territory of our everyday life as a temporal stratification, a sediment, rather than as the effect, of policies, techniques, cultures, symbologies and representations, behaviors and social norms, cooperation and conflicts. At the same time, a territory is also the outcome of administrative and design practices, with their multiplicity of morphogenetic elements that interact with each other and with the physical territory. This way of studying a territory is enriched by its consideration of a multiplicity of vocabularies, taxonomies and ways of observing and listening to the territory.

Looking at the Po River Delta in a way that is plural and multiple seems the most effective way to spare this territory from being examined in a manner that's as precise and deep as it is specialised and unable to reveal the sometimes fragile but important connections between different processes and phenomena. For this reason, in our research we adopted techniques and methods from different disciplines: anthropology, visual communication, planning, urban design and landscape architecture. Observing the Po River Delta from this multiplicity of perspectives was the starting point and foundation for arguments strong enough to allow us to explore different scenarios.

The investigations and scenarios collected in this part of the book aim to add to the knowledge about the Po River Delta territory, by contributing a fragment that seeks to put data in relationship and in context, avoiding the segmentation and self-referentiality that often characterize specialised design proposals.

Explorations in the Po Delta

Enrico Anguillari

Emanuela Bonini Lessing

Marco Ranzato

Introduction
Enrico Anguillari
Emanuela Bonini Lessing

The implicit assumption of the entire study is that the territory itself should be regarded as a player and, as such, have the right to speak: it should be heard and not just described, it should be allowed to express its needs, to participate in its own transformation, to evaluate its ability both to react and to act. Instead of adding to or overlapping the complex skein of players and local interests, or devising additional sector plans or specific projects that tend to embrace one point of view at the expense of others, our research activity has above all attempted to help the local community become aware of its own weaknesses and opportunities. All this, with the intention of finding points of agreement concerning the expectations and the different ideas about transforming the territory for the future.

To do this, we thought it appropriate to unite the respective expertise of planners and of communication designers, both in the early stages of defining the research topic and during the more active stages in the field. Concretely, these two disciplines complement each other during the phases of the creation of ideas or visions of the territory, and of the communication of these ideas to and dialogue with the local population. The various instruments - consisting of communicative artefacts on one side and scenarios on the other - create a common corpus capable of soliciting a collective reflection from the inhabitants, based on planning ideas. The goal is to "force them" to form an opinion and adopt a critical point of view on specific issues concerning the various futures of the territory they inhabit.

Our study also represented an important opportunity to highlight the multiplicity of roles that communication designers can play in designing a territory, especially when the latter is a matter of public interest. That is, the communication designer can be "just" an individual who attentively translates into graphic form decisions and expectations previously expressed by others (other professionals, local stakeholders), or this individual can help shape the entire planning process. In most cases, in current design practice, the designer intervenes when the work of defining the identity - one could almost say creating an identikit - of what he or she will communicate graphically has already been completed at an earlier stage, to which he or she does not always have access. Instead, in the case of the research on the Po delta, the final objectives, methodology and research tools to achieve them were discussed and shared with urban planners, site planners and landscape architects. In fact, the work of studying the territorial identity and of surveying the population involved rethinking the aims of the communication project, the role of the designer, and the figure of the commissioning party and of the final users of the project.

On the one hand, the visual designers' job was to have a grassroots identity of the territory emerge, and then find a way to give a visual representation to it, corresponding to the residents' perceptions of their living environment and to their expectations about the future. On the other hand, the scenario building process, which was carried out parallel to the process more strictly linked to visual communication, was meant to be a practice aimed at researching, discussing and sharing a vision of the future for the area concerned.

Another aspect that played an important role in defining the topics and the manner of research, was the careful collection of case studies relating to research and design, and to good practices that had been used in other territories similar to, or characterized by dynamics analogous to those of, the Po Delta. This made it possible in each situation to think about the scenario's role, and the choices made for such scenario with their relative consequences, on the basis of reflections, discussions and sometimes direct testimony.

In this work, therefore, the scenario is not intended to be a long-term program or set of programs for the territory, aimed at making future realities fit into a previously defined framework. Nor is it the result of the "design of a Design" (Ascher, 1995). Rather, it should be understood as a set of tools whose use reveals the potential and constraints imposed by the society, places, circumstances and events. Compared to a plan or project for the territory, the scenario is based on more reflective guidelines, suitable for an uncertain future, and articulates a consistent and diversified "coming and going", the long-term and the short-term, the large-scale and small, the general interests and the special interests. In this sense, the practice of scenario-building is like that of strategic territorial management, which takes into account the territorial expectations, uncertainties and events of an open society.

Starting from these considerations, the scenarios presented in these pages take as their temporal horizon a sufficiently distant future - 2100 - to allow it to test the limits of the delta territory's fragilities and to forge some hypotheses. To emphasize the exploratory nature of the scenario, the question "What if?" is asked. For example, what if the average sea level rises 1.5 meters in the next 100 years? What if the financial resources to ensure the territory's safety were lacking? What if the soil's salinisation made the land unproductive?

Just as these questions seek to generate a reflection on a common future (WCED, 1987), in similar fashion, the territory should be seen as common (Nagendra, Olstrom, 2008), that is, a common good which whose essentially cooperative form of ownership and management allows the property to be used by anyone who wants to, following rules to ensure that such uses can be infinitely reproduced.

References

- Ascher F., (1995), *Métapolis ou l'Avenir des villes*, éditions Odile Jacob
- WCED (World Commission on Environment and Development), (1987), *Our Common Future*, Oxford University Press
- Nagendra H., Ostrom E., (2008), "Governing the commons in the new millennium: A diversity of institutions for natural resource management", in *Encyclopedia of Earth*, Cleveland, Washington DC.

Notes on the visual identity of the territory
Emanuela Bonini Lessing

Once the aims of the work had been shared with the urban designers, we tried to formulate more precisely how to conduct the research into the territory's identity, starting from the methodological and more strictly operational tools of visual communication design.

The somewhat abstract question formulated at the beginning, "How could we visually represent this territory?", soon gave way to another question, more appropriate to local conditions, namely, "How would you like your delta to be?"

The fundamental difference between the two is that the first question focuses on how the territory's identity could be represented by the designers. The second, instead, asks what are the characteristics of that identity, that is, if and how the residents perceive themselves as a community with a past and a future that unites them.

This question helped reshape the entire communication process, putting the residents at the heart of the matter. Since one of the aims of the research into the Po delta was to raise the level of awareness and knowledge on the part of the local community concerning the development opportunities for the future, two steps immediately proved necessary. First, to find a way to listen to the residents, to their expectations, their perceptions of the current situation, their personal and collective memories. To reconstruct, in a sense, the "personality of the place." Second, to enable people to express their thoughts and opinions on the results of this survey, and on the development scenarios proposed. More than trying to inform the public about the project activities underway and its objectives, we tried to involve them as much as possible in the process of setting up the research itself. The designer's task was to identify and propose elements - which would also be visually useful - in which the citizens could recognize their territory and could discuss it, to then make pondered decisions. In a territory as dispersed and fragmented as that of the Po Delta, the circula-

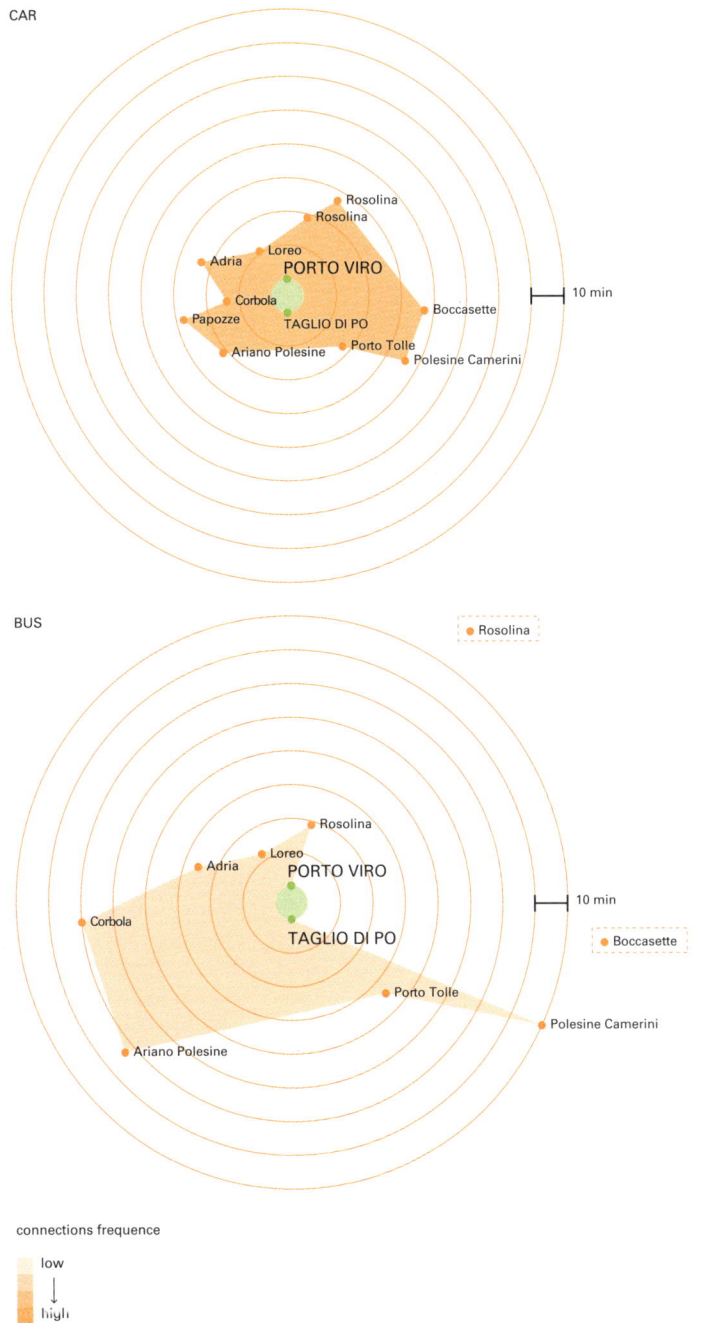

ATLAS - Second Part | Po Delta | Explorations in the Po Delta | Emanuela Bonini Lessing

DIVERTIMENTO   SILENZIO   TRASAPARENZA
  FINE
        FREDDO   SPAVENTOSO   ORIZZONTI
  FRAGILE   INSTABILE   ADRENALINA   PESCA
UTILE INCONTRO   GENEROSA   SCHIUMA
          RELAX   VELOCITÀ
SICUREZZA   LETTO   TRISTEZZA
           PALUDE   CASCATA
    DUTTILE   SQUALI   MARGINALE
   PERICOLO   CONTAMINAZIONE
SUOLO   POTENZA   LIMITE
  DINAMICO   ZOLLE   ROSSO   ACQUA SPORCA
       INFINITO   VEGETAZIONE MOLTEPLICE
SFIDA   PIANTE
     VERDE   ASSESTAMENTO   SOLITUDINE
ARATURA
    INTRECCI   TERRA BATTUTA   BEATITUDINE
CIBO   RISORSA   FERTILE
    ANNEGARE   DIFFUSO   ALBA
CONFINE   SPAZIO   CORRENTE   CAMPAGNA
      CAMMINARE   MONTAGNA
ULISSE   VERDE   MEDUSE   ARIDA
  RURALE   NINFEE   VASTITÀ   CANNETO
     COLTURE   AVVENTURA   CAMPO ESTESO
NATURA   COLORI TENUI   TRADIZIONE   PERCORRIBILE
           VITA   ARCANO   POTENZIALITÀ
VENTO   SCUOLA ELEMENTARE
    DELINEATO   NOSTALGIA   SCORRERE
COLORE   FENICOTTERI
      STRADA   MARGINE   RADICI
    TANTISSIMA PAURA   SERPENTE   SC
SOLE RETE   RICCHEZZA
  IMMENSA   CALDO   BLU
         ODORE SALMASTRO
    ARRABBIATO   INSETTI
         SCOMODO   EVOLVERE
                       ABBRACCIO AL MARE
  ODORE DI PIOGGIA
  DOLCE   CALPESTARE
       MARTORIATO   VITA
TRADIZIONE   ARGINE   EQUILIBRIO   DOMINABILE
    OMBRELLONI   MESCOLANZA
SOLIDITÀ   ALLEGRIA   CULTURA
      NEBBIA   COMPLESSITÀ
ORGANISMI   AGRICOLTURA   FLUSSO LENTO
   DEVASTAZIONE   ISOLAMENTO
              OMOGENEITÀ
FRUSCÌO   ASSENZA   CANNETI   CAPELLI CRESPI   EQUILIBRIO PRECARIO
              SALATO   RICCA
  ZOLLE   NON ARTIFICIALITÀ
SCIABORDÌO   INGOVERNABILE   FIORITURA
     PROFUMO   IMPREVEDIBILE
FRAGILE   MISTERIOSO   LETTERA GRECA
   AZZURRO CASA   VITA
VUOTO   SOFFERENZA   SPAZIO APERTO   LABIRINTICO
TRAMONTO   LIBERTÀ   RELAZIONI
   ZANZARE   VOLATILE   DINAMISMO
EGIZIANI   PROFUMI   TRANQUILLITÀ   IBRIDO
CIBO   NODOSO   BARCA
     MORTE   EMERGENZA
BELLO   PESCA   INQUINAMENTO
   DISPERSIONE   TRIANGOLO
          DISTESO   RAMI   ENERGIA
FLUIDITÀ   ENORME
     IMMANENTE   PIENO

tion and sharing of information are crucial issues. Thus, a first level of objectives pursued was entirely internal, one might almost say for the sole benefit of the citizens of the delta.

What characterizes this area in its inhabitants' perception? And then, what interventions, devices or services would help improve the quality of life of those who live here? To answer these questions raised by this same study, it was essential to meet the residents, interview them, film them, and take photographs. A search was done into the archives of the local newspapers, which mostly describe these lands as still suffering from the phenomena of emigration and poverty after World War II. Bibliographical research was undertaken, and photographical archives were consulted, which instead also showed the territory's positive connotations. Visiting the museums for local culture, it was pointed out to citizens how much of today's industrial production continues to draw upon the experiences and traditions of this place. The survey work was joined little by little by the work of provoking and stimulating dialogue and joint reflection on the local cultural heritage, such as when large posters were specially made with quotes from novels by famous authors from the delta, posted on the walls of houses and farms. During the weeks of the research group's stay at the Po delta, public presentations and debates were organized, in which the public as well as the main representatives of institutions and local associations were invited to participate. The overall objective was to activate a kind of "mirroring" process, that is, metaphorically giving citizens a "mirror" that could reflect some common features of the area, on which they could then engage in dialogue. As the fieldwork progressed, a dual reading of the area emerged clearly. On the one hand, the work made evident the fragility and the fragmentation afflicting the people's sense of belonging to their territory, which is marked by a deeply-rooted pessimism. On the other hand, there emerged complex but possible opportunities to combine the various economic and social vocations of the area and to allow them to coexist; until then the residents had perceived these opportunities only scarcely or not at all.

To meet fully the objectives of our research and work of proposing scenarios, it soon became clear that we couldn't obtain a complete picture of the situation without considering the external players in the territory. To what extent is it possible to discuss the welfare of the local population if we exclude its relations with the outside world? It is only by opening up the territory and its services to the rest of the world that the development process can be triggered. However, it is precisely when the local population is little aware of the resources and tools to configure their own future, that the presence of external actors, the expression of strong political and economic powers, can engender further crisis factors. The analysis, conducted also by using the theoretical tools of visual anthropology, made it possible to construct a map of the powers operating in the area, not so much from an objective point of view as from that of the residents. Consistent with the idea of keeping the local community at the centre of the project, we tried to depict the local community's perception of the distribution of power.

There emerged an intricate system of relations among local actors, aimed at achieving goals which today necessarily involve denying opportunities to some in order for financial gain to accrue to others. The separation of interests between social groups present in the delta is visually represented, for example, also by the totally different use that the farmers and clam growers make of the territory: two activities whose opposing economic revenues, times and places make them completely unrelated to each other. Though sometimes there arise unexpected - but misleading - instances of cooperation between actors, who against their will find themselves sharing common interests. Faced with an obvious difficulty on the part of the local government to carry out certain maintenance tasks for the territory, other private entities have offered to do so, with the real goal of strengthening consensus in their favour, thus spoiling the relations between all the local actors involved.

The total result is that the current delta is the product of zonings and settlements, depending

01 *Time to travel maps*: the perception of the territory's spatial dimension changes according to the means of transport used; the more the slower the means of transport and the less frequent the connections, the larger the territory is perceived to be.

02 *Landscapes of words:* Two points of view on the Po delta: that of the inhabitants (left) and that of the students involved in researching it, who visited the area for the first time (right)

03 *Perception of power distribution:* The diagram represents the perception of the delta's inhabitants concerning the distribution of power. The Veneto Regional Park of the Po Delta would have almost no influence on land management

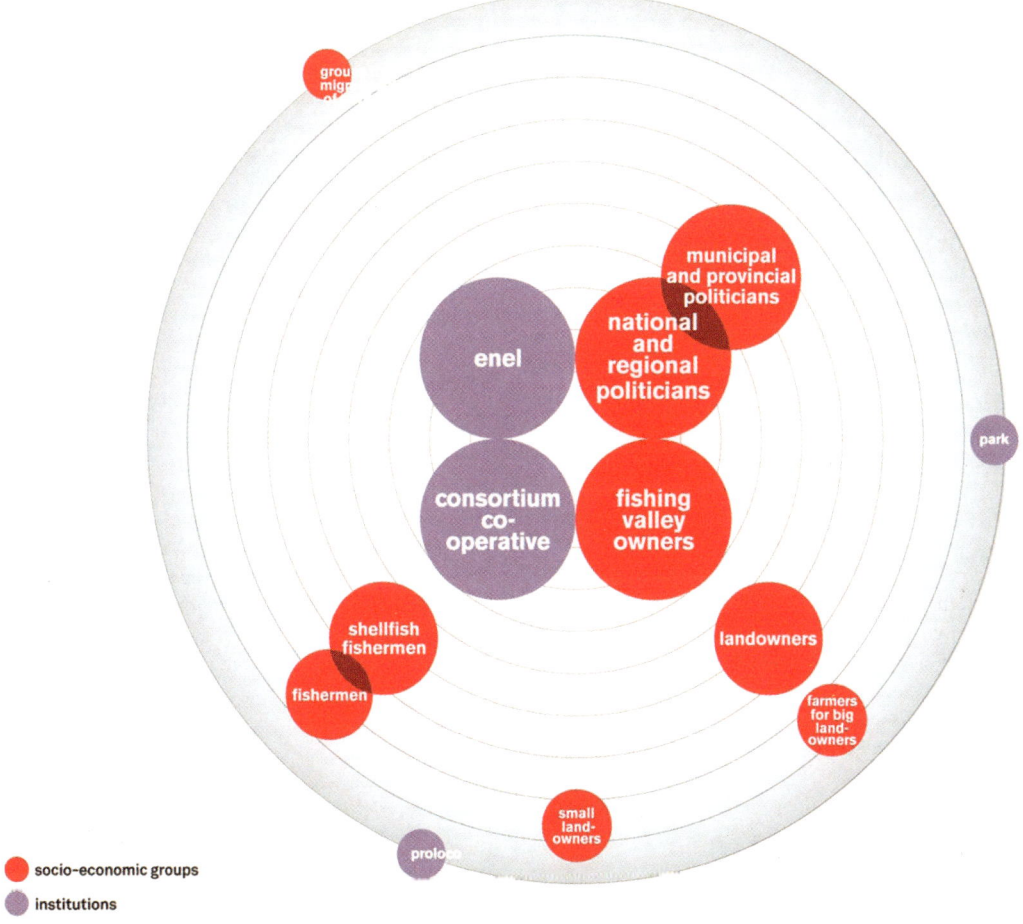

04 *perceived problems*. All these issues were described by the interviewed locals as problematic to their future and that of the territory in general.

Power plant pollution
Heating lagoon water
The factory pours hot water into the sea to cool down the machines.
White coal pollution
The boats carrying white coal to the factory are (illegally) washed in the sea.
Cancer

Illegal shellfish fishing
High degree of conflictuality among fishermen (low social capital)
Some people earn more than others thanks to illegal activities provoking their neighbours' jealousy; it is impossible to denounce for fear of retorsions.

Hunters in fishing valleys
There is a lack of regulations inside the fishing valleys; as a result, rich hunters kill an enormous and indiscriminate quantity of birds.

River pollution
Eels do not hybernate in the bottom of the river anymore because the sediments that the river brings from the north of Italy are too polluted.

Sediments accumulation
Accumulation of river sediments at the entrance of the lagoons ("sacche"); the Province, the institution that should excavate the sediments, is not doing its duty.

Diminishing number of fish species in the river
The 'pesce siluro', which was introduced by mistake in the Po river, has eaten most of the fish species in the river, and is now its main inhabitant.

Reduced mobility
Children who live in dispersed areas have difficulties in getting to school.

Role of 'political godfathers' in the cooperative system
In order to get access to licenses and permits you need to have some political gofather ("santolo") who is going to promote your candidature. It is impossible to transmit one's permits directly to family members.

The park is not bringing any improvement to the territory
There are no clear consequences – either negative or positive – of the implementation of the park in the area; the park authorities earn money without doing any good to the territory. They have no power to influence other decisions concerning the territory (for example, the power plant).

Depopulation
Young people emigrate in search for new opportunities.

FISHERMAN

FARMERS

COOPERATIVES

PRO-LOCO

STUDENTS/ LIBRARIANS

on which interests have prevailed in each area, an archipelago of juxtaposed islands, each with a strong specificity and autonomy. This dynamic continues to fuel the separation between the players, concerning the organizational strategies, economies, cultural and social processes, and how to manage the territory. Unfortunately, in a selfish power play, external economic powers take advantage of the opportunities for "conquest" offered by the delta's more or less objective fragilities, with the intention of exploiting the territory rather than nurturing its coordination and recovery. Both the individuals and the economic players are always on the lookout for anything that might reassure and establish confidence: their willingness even to embrace measures with only temporary effects further exacerbates internal conflict.

This contrast - together with objective environmental problems - causes the local population to perceive the territory as fragile. It does not offer a socio-economic structure or a structure of government that foster a sense of belonging to the territory on the part of its citizens.

This fact is by no means trivial. No project to transform the territory can begin and hope to reach completion if the population is not motivated by a strong sense of pride. Pride in belonging to that specific territory, which values its natural, social, economic and symbolic characteristics as core qualities making it distinct from the rest of the region. A sort of "territorial pride" that does not mean contempt of the Other, but a stimulus to undertake the necessary changes to improve their relationship with the territory.

The study further highlighted how the conflict between players and between different expectations translates into opportunity or the denial to access portions of the territory. Concretely, this means the impossibility for the majority of the population to benefit from large areas, the resources contained in them and possible alternative economic and social functions with respect to those presently in place .

All these factors have allowed us to detect the "presence of a great absence", that is, the entity commissioning the project for renewal and development of the delta area.

While the University played an important role in indicating a methodology to address local issues and has produced over time a substantial amount of material for the local players to examine together, it was either unable or unwilling to grasp the opportunity presented by the mutual comparison, and didn't follow up the study with concrete actions. The fragmentation of interests and economic and political powers, so clearly evidenced and represented with the help of the visual design, has been from the beginning of the project an irremediable fragility. The fundamental role of the authority commissioning the study cannot be limited to expressing an unease and voicing the need to deal with it: it must welcome the expectations and needs of all local representatives, not just of one faction. It must support the designers in their work, making it possible for all the other stakeholders to dialogue, exchange viewpoints, and test research and project hypotheses. If not, even the positive and innovative research efforts, despite the client's best efforts, will end up being interpreted as responding to the requirements of only some parties at the expense of others, risking to delegitimize the entire scientific proposal.

The study has attempted to compensate in some way for the absence of this important figure, acknowledging the resident community simultaneously as both recipient of the project and, at least in part, the commissioning party. This at least legitimized an unprecedented process of investigation and representation of the identity of the place and of the local community. A process of highlighting and sharing many fragments, sometimes scattered, hidden or flaunted or which domineer over others. If we arrived at the Po Delta with the idea of constructing a unitary and coherent image of the territory, we left with an understanding of a complex reality, unable, at present, to give itself a shared visual representation.

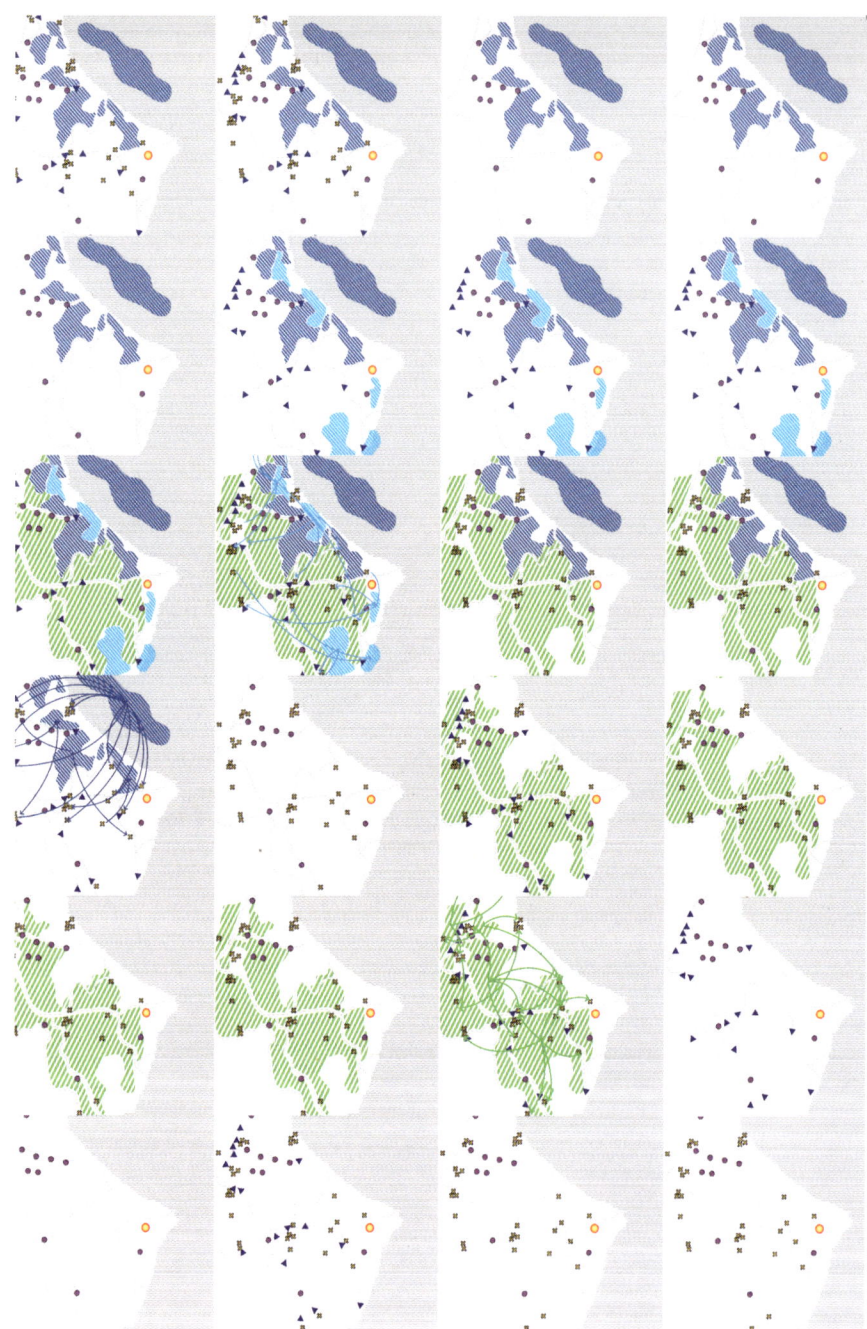

05 *A day in the delta_a* the main economic activities are concentrated in specific areas of the territory, each used almost exclusively by one specific type of worker. The maps represent land use over 24 hours, starting at midnight (the first box in the upper left corner)

06 *Sound identity* at first the delta seems empty: at a glance, everything seems flat, human activities inexistent. Then, a slight buzz emerges: the faint sound of nature and work activities scattered throughout the territory.

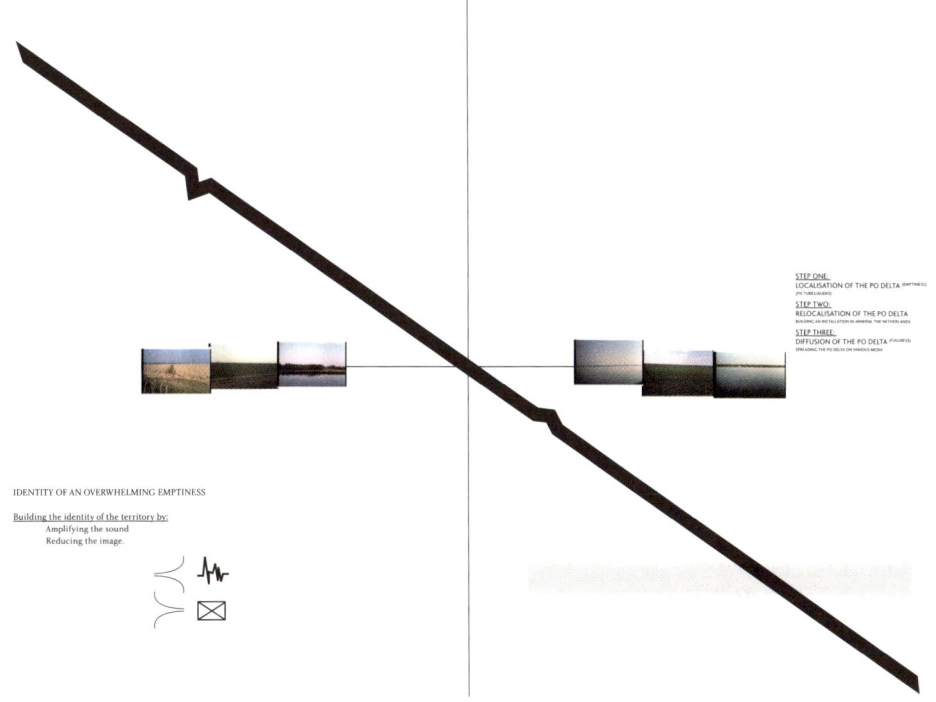

Po Delta 2100
Morphogenesis of a cultural landscape
Enrico Anguillari

As we witness an inexorable process of global urbanization, the territories we live in are becoming increasingly fragile and vulnerable to the impact of climate change.

The Po River delta, in the Veneto region, is no exception. Here, the long work done to build a habitable area is now being challenged by a number of environmental problems and serious disorders facing the local population.

In every place a close relationship can be perceived between man and nature, or rather between man and water, which results in a strong 'sense of belonging' to an area reclaimed from the sea with great effort. The people of the delta describe 'reclamation' as the daily effort necessary to keep the area safe, habitable and productive.

The delta's balance is weakened also by everything that is poured into the whole catchment area of the Po River and drained downstream. Most of the environmental problems the inhabitants of this area have to face are thus the result of both exogenous and endogenous dynamics.

The pollution of the water and soil, eutrophication, flood risk, saltwater intrusion, salinisation of the soil, scarcity of fresh water, and subsidence highlight how the river itself is no longer able to activate its own forms of response to compensate for these dysfunctions. In the summer, due to the reduction of its discharge, the sea water is pushed up the branches of the river as much as twenty kilometres from the mouth, causing serious problems for the settlements and the entire agricultural economy. Due to the upstream sediment retention, the coast is gradually eroding, exposing the entire area to the risk of storm surges (link imm. Weaknesses).

In any case, although strictly harnessed, the delta continues to be a morphogenetic driver: its lagoons continue to be fed by the sea, and its land is periodically flooded and drained. The sediments carried by the river continue to settle along the coast, which continues to be shaped by the wind and tides (D'Alpaos, 2009). Therefore, because of its intrinsic characteristics, it cannot be considered a stable territory whose issues have been resolved; rather, it should be viewed as a dynamic and animated phenomenon, characterized by a set of composite structures in continuous movement (Tosi, 2011). Its precious ecosystem provides excellent conditions for the development of specific economic activities. This is a heritage that can support and enhance the local identity, promoting tourism, fishing, aquaculture and high-quality agricultural production (link imm. strengths).

However, it remains necessary to ask whether, and for how long, this area will be able to cope with the dysfunction to which it is currently exposed and whether it will be able to adapt to events that climate change promises.

Given that, in a hundred years, the predictions about the growth of the mean sea level are around a meter and a half (Bondesan et al., 1995), what implications does this have for the environment, economy and society?

Trying to imagine mechanisms by which and through which local structures of the delta will still be able to 'fight' or 'adapt' to the pressures upon it, requires a thorough knowledge of the ongoing dynamics and their effects on the places and ways of life, and obliges one to rethink the way in which the fragilities and land resources could be addressed in order to define new visions for the future. The hypothesis presented here offers a vision for 2100.

Two antithetical scenarios clarify which strategies it would be necessary to adopt if one wished to cope with existing environmental pressures by opposing them - in the idea of a resistant, strong and infrastructured territory - or by absorbing them - in the idea of a resilient territory, adaptable and capable of transforming the fragilities into potentials. What advantages or disadvantages would the Po delta face in either case?

The delta is the result of natural processes and human interventions aimed at controlling them.

The first scenario aims to oppose the threats of climate change by a greater strengthening of the territory and by increasing its infrastructure (link imm. counteracting). It will be necessary, in order to ensure water safety and maintain a stable landscape in anticipation of a meter and a half increase in the mean sea level, to plan interventions in order to obtain the maximum benefit in terms of risk reduction and of a achieving minimal impact on the local identity, culture and way of life of its inhabitants (Ranzato, 2012).

The coastal dunes should be protected by embankments and dykes to keep the sea from eroding them and to allow the river to continue to build the coastline - coming to form 'superdunes'. Even the embankments separating the fish farms from the lagoons, as well as those along the branches of the river, will have to be reinforced, extended and raised.

The only 'escape route', their summit, could become a vantage point on the landscape and a real 'multifunctional space' rich in facilities for leisure, tourism and the various activities characterizing the delta. Together, the embankments will act as a territorial frame able to connect places of exception, natural resources and cultural heritage that are spread out from one another. In addition to ensuring the territory's safety, these 'spaces' thus would contribute to improving the accessibility, use and knowledge of the area.

To reduce the river's flow during floods, a series of 'rooms' and expansion basins would be dug, which can also be used to hold fresh water and irrigate farmland in times of drought. The removed soil would be used to raise and increase the embankments.

Within the reservoirs, the planting of phytoremediative plants will allow for a containment of eutrophication by reducing the excess nutrients and fertilizers given off by the crops that, in lowering the level of oxygen in the water, now threaten to make large portions of land unsuitable for other species - for instance, for fish.

Some basins can accommodate photovoltaic systems mounted on floating platforms; these would contribute to the system's energy self-sufficiency without taking up space for crops, and would slow the evaporation of the basins themselves.

- urban settlements
- railway
- romea road
- extensive agriculture
- intensive agriculture
- existing dikes
- reinforcement of dikes
- wave brakers
- superdunes
- existing irrigation system
- connection channels
- water storage basins
- wave brakers/superdunes
- research laboratories

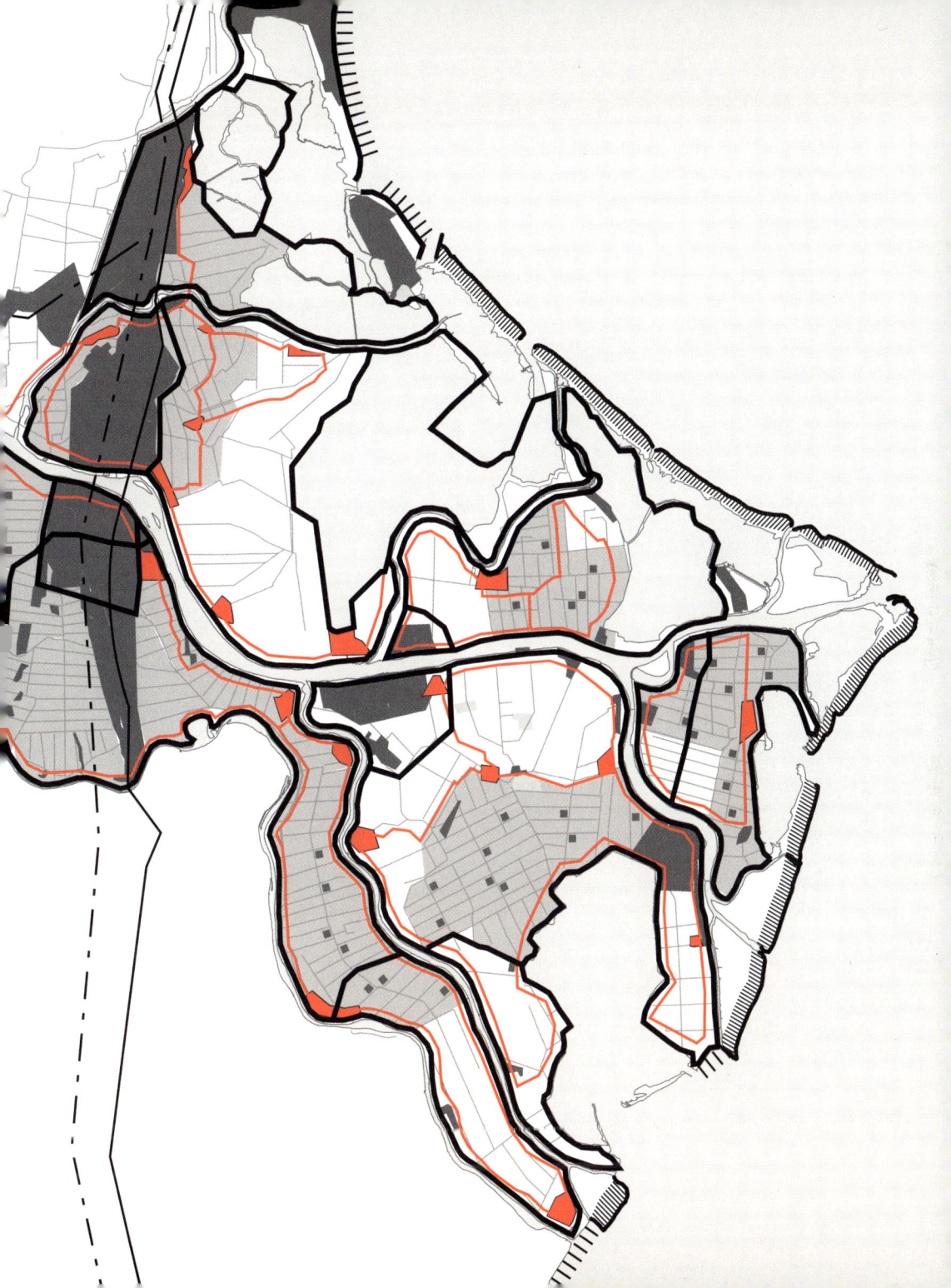

Thus, at the foot of the embankments a sequence of environments typical of the river landscape - riparian forests, meadows, wetlands and meanders - would develop, alternating with large bodies of water and 'floating solar farms'.

Water, nature and energy will be pushed into the most remote places and will innervate the territory following the routes of the drainpipes and farm roads. Taken together, these ecological/technological corridors would build an 'activation network' able to support and strengthen the different landscapes and different economic activities of the delta.

Increasing the territory's safety will also improve its overall habitability. The settlements located along the ancient barrier islands will grow, giving the area a 'more urban' feel, similar to a city overlooking a large park.

The society will head in the direction of a 'knowledge society' and the delta will become a laboratory for experimenting with new farming and aquaculture techniques. This will allow for a transition from the present large-scale farming to small-scale, high quality production and diversified family economies.

Improved accessibility to the area, driven by the main territorial framework, will encourage the population to go back to living in the countryside - for example, in settlements built during the agrarian reform period -, to organize themselves into cooperatives and take care of their territory.

Abandoning a strictly technical approach, nature and society must be treated as complex structures, unfit for automatic command and control mechanisms (Mostafavi et al., 2010).

Accommodating the changes due to climate, then, means questioning the cost - including the social cost - of maintaining and strengthening individual structures, networks or large portions of the territory, and means bearing the consequences of the profound hydrogeological changes that would set into motion the entire delta landscape (Anguillari, 2012).

In 2100, the increase of the mean sea level will make the eventual floods harsher, inundating

- urban settlements
- reiforced dikes
- reiforced dikes with paths
- shadow dikes
- wave brakers
- sea-lagoon
- fish farms
- forests
- swamps-fresh water basins
- irrigation canals

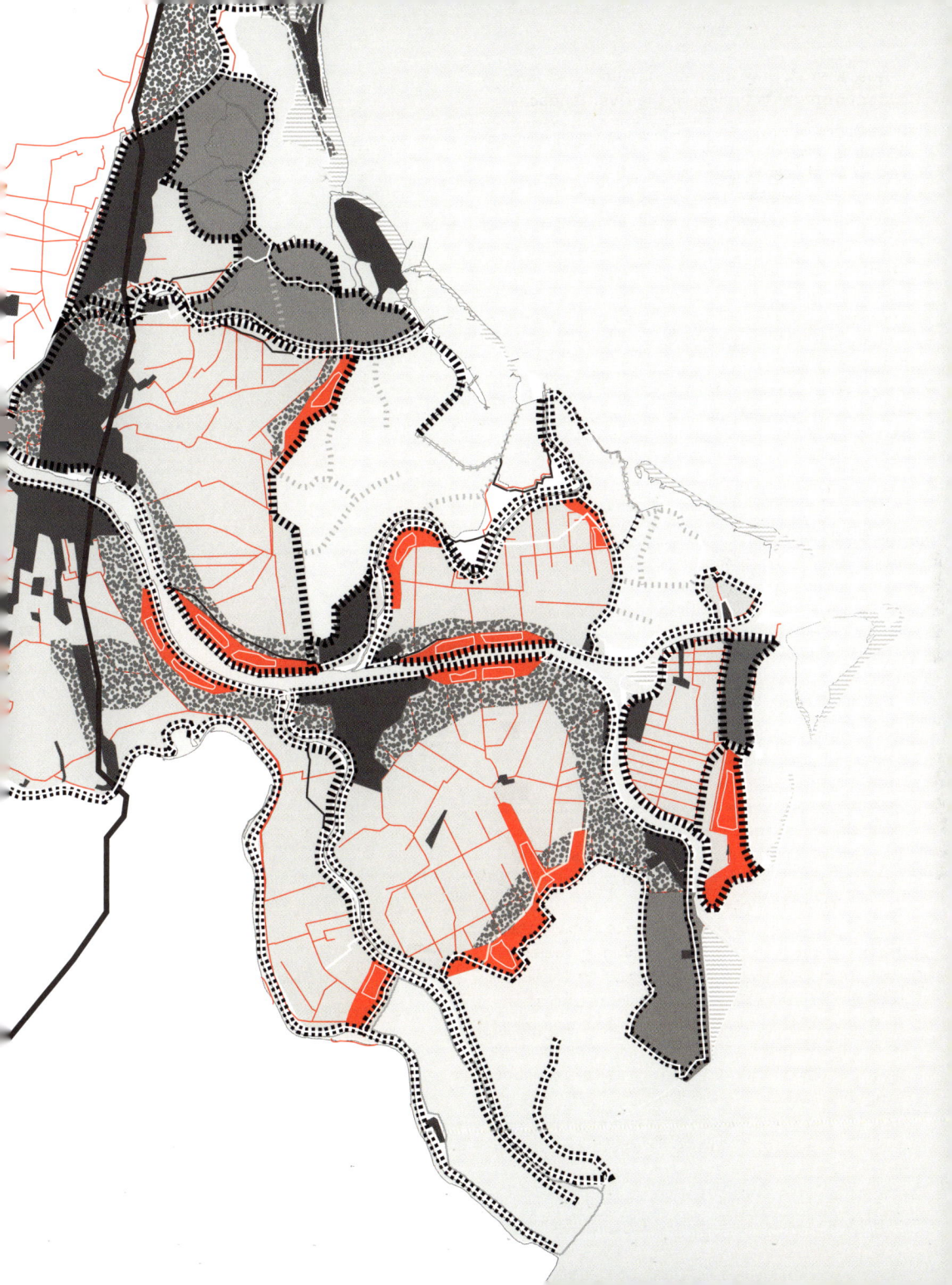

TIMELINES

2011

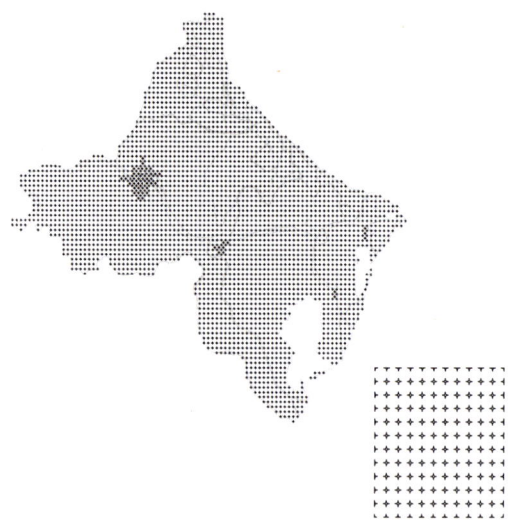

society		preserve landscape and cultural identity		encourage the population to settle in the densest areas	promote a philosophy of positive adaptation to environmental changes	encourage support networks for communities to face changes
			encourage maintenance of existing buildings			
economic resources			prioritize fish farming	promote agrotourism	continue pumping water at current rate to sustain agriculture and fish farming	prioritize shell-fish farming
				prioritize agriculture		
2011	2020		2030	2040	2050	
			prioritize new forests to counteract future erosion		prioritize new wetlands	
			retain characteristics of the northern area	create wetlands up-stream to clean industry pollution	allow more flooding from the sea	
environment						
		encourage solar energy generation			adapt infrastructure to facilitate proposed changes	
infrastructure		create new basins to store fresh water against salinisation				

Enrico Anguillari | Explorations in the Po Delta | Po Delta | Second Part - ATLAS

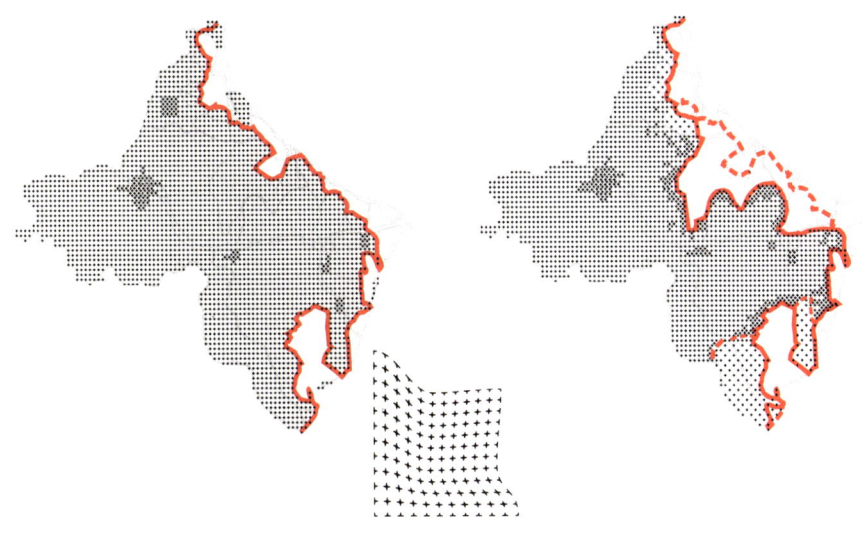

ATLAS - Second Part | Po Delta | Explorations in the Po Delta | Enrico Anguillari

○ main centers
° touristic ports
· commercial ports
✗ regional connection
— dikes
▨ productive areas
▦ shellfish farming
▦ hunting
▤ protected areas
≡ drenage system

▨ salinization
⊃) subsidence
||| fragile border
✓ emigration
▨ fish farming
    fishing
⋮⋮ agriculture
══ main roads
— railway
▪ waterpumps
● small centers

Enrico Anguillari | Explorations in the Po Delta | Po Delta | Second Part - ATLAS

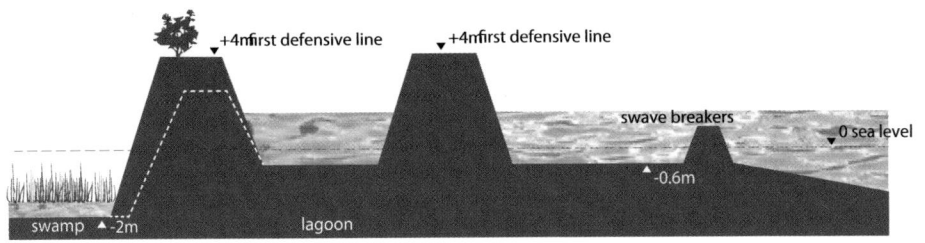

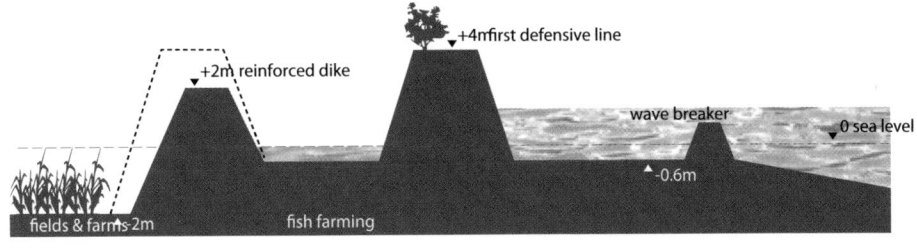
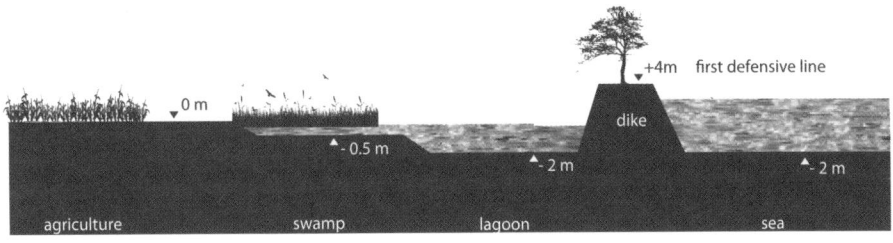
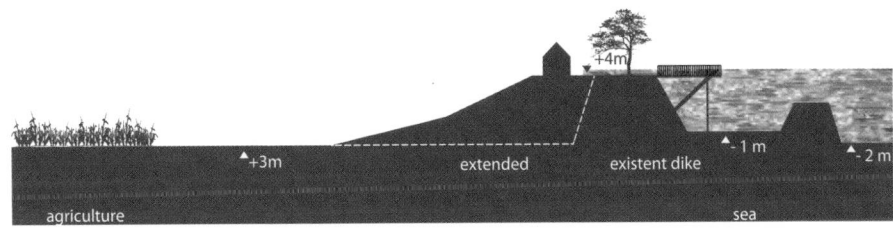

ATLAS - Second Part | Po Delta | Explorations in the Po Delta | Enrico Anguillari

large portions of the territory and pushing back the current line of defence, gradually eroding dunes, dikes and dams (link imm. accommodating; link imm. timeline).

Fish farms, now strictly regulated, will become transitional environments in continuity with the lagoons or with the bays open to the sea.

In contrast, the river will take back its own spaces, creating wetlands, marshes and swamps with greater vigour. In some cases, by taking advantage of the land's contours, these will become natural reservoirs of fresh water used to irrigate the fields. The strong hydraulic pressure put on the embankments by the river's mass, will help reduce the infiltration of salt water into the soil. Through controlled flooding, the lower areas will be filled by sediment deposition, compensating for subsidence, compacting the soil and reducing the level of salinity.

Despite a general improvement in the quality of the land, it will still be necessary to introduce new forms of agriculture and more adaptable crops that require less water than the current ones.

To keep the cost of managing the network down and to minimize water leakage, it will be necessary to close the system loop of 'collection-phytoremediation-storage-reuse' and to power it using solar energy or wind power, making the individual islands 'autonomous' (link imm. masterplan Porto Viro + hydraulic diagram).

As already mentioned, the cultural landscape of this area is the product of a geo-socio-historical context characterized by strong tension between natural processes and human intervention.

In the idea of a resilient territory, many of its characteristics will be lost, if left to the destructive effect of climate change. On the other hand, there will be a greater richness in terms of biodiversity and environmental quality which, if well managed, will become a resource for diversifying local economies by increasing their complexity.

It will therefore be necessary to select and re-insert buildings, infrastructures and contexts into a different cycle of use, contributing to the composition of a new multi-dimensional landscape (link imm. sections). This is the case of the embankments, the water pumps, the settlement of agrarian reform, the manor houses, the buildings of industrial archaeology, the fossil dunes, the relics of marshes, woods and individual trees. But also of territorial matrices that still generate peculiar economic activities that are easily integrated into the circuit of high quality products and of tourism - such as fish farms, aquaculture, horticulture, rice cultivation. That is, a set of elements that, by strengthening the territorial identity and the local population's sense of belonging, can still be the cornerstones of an agenda aimed at enhancing the socio-economic context of the delta to come.

References

- Anguillari E., 2012, *Accommodating environmental pressures*, in Tosi M. C., Anguillari E., Bonini Lessing E., Ranzato M., (eds), Delta landscape 2100, Trento, professionaldreamers, 31-53.
- François Ascher, (2001). *Les nouveaux principes de l'urbanisme*, Editions de l'Aube.
- Bondesan M., Castiglioni G.B., Elmi C., Gabbianelli G., Marocco R., Pirazzoli P.A. and Tomasin A., 1995. *Coastal areas at risk from storm surges and sea-level rise in Northeastern Italy.* Journal of Coastal Research, 11, 4, 1354-1379.
- D'Alpaos L., 2009, *Evoluzione morfologica recente della Sacca degli Scardovari*, Quaderni di Ca' Vendramin, 0, 16-52
- Mostafavi M., Doherty G., Harvard University Graduate School of Design (eds), (2010). *Ecological Urbanism*, Baden, Lars Müller Publishers.
- Ranzato M., 2012, *Counteracting environmental pressures*, in Tosi M. C. et al., (cit.), 55-77.
- Tosi M. C., 2011, *Living the park: the need of a conceptual shift*, in Tosi M. C., Anguillari E., Bonini Lessing E., Ranzato M., (eds), Delta landscapes. Geographies, Scenarios, Identities, Rijswijk, Papiroz Publishing House, 7.

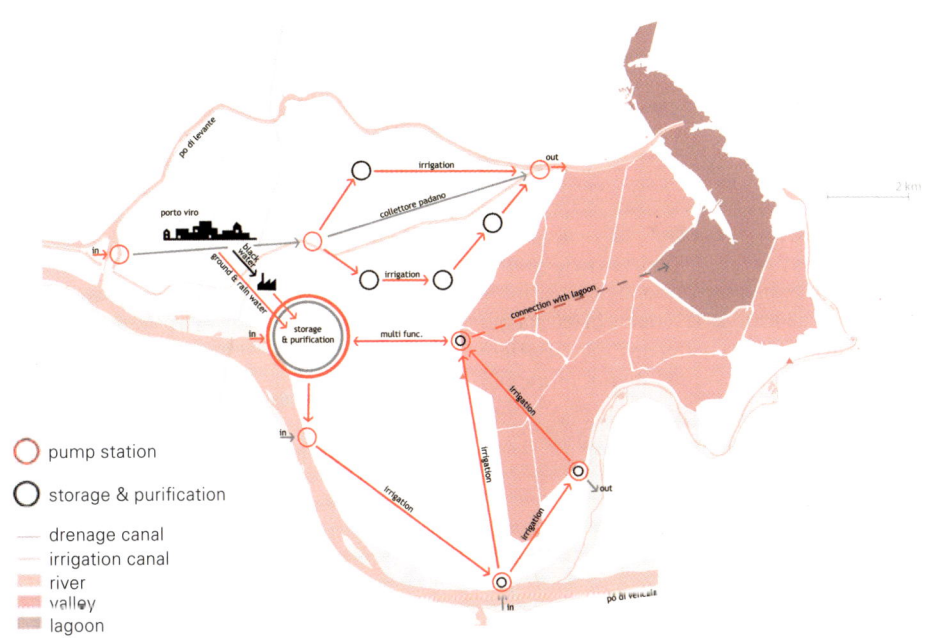

ATLAS - Second Part | Po Delta | Explorations in the Po Delta | Enrico Anguillari   137

Actions for a deltaic resilience
Marco Ranzato

◇◇◇◇◇◇◇◇◇◇◇◇◇◇◇◇◇◇◇◇◇◇◇◇◇◇◇◇◇◇◇

In recent years, development strategies and projects have been considering a different approach to organising and developing deltaic territories. The boundless confidence in engineering equipment that allowed uncertain terrains between land and water to be conquered has in many cases faltered due to the many signs of fragility, if not of collapse, of the engineered.

Overflow and floods, periods of prolonged drought, salinization, 'natural' lowering of the terrain (subsidence) are issues that albeit with different intensities, have affected the deltas across the board.

New strategies and projects consider the delta a living territory where different and intertwining forces and materials can be seen. Often new strategies and projects are compared with the short and long term climatic forecasts. Even more often, they do not answer matters in just one direction (solely water-related answers, for example) but consider a variety of issues which sometimes are integrated with visions involving the deltaic environment as a whole. According to the definition of resilience provided by the ecologists van Leuwen (1972) and van Wirdum (1982), strategies and projects can be interpreted and measured based on the ratio of resistance (resistance as counteracting) or of retention (retention as accommodating) that they establish with the environmental forces and dynamics proper to deltas and estuaries. According to this ecological perspective, the overall balance between each part's resistance and retention determines the system's capacity for resilience and therefore the delta's degree of resilience. Another relevant aspect for reading such strategies and projects is their ability to establish synergies with the environmental dynamics, that is, the degree to which the physical transformations they imply are integrated with the overall deltaic environmental processes - understanding the anthropic processes as part of the environment. The basic assumption is that the delta's dynamics are alive and are asking for projects that establish the befitting interplay with them. This is of ultimate importance to build up resilient deltas. In the bargain, as McHarg (1967) realized through a careful reading of the environmental dynamics, resilient solutions can be found just in the exact interpretation of the environmental processes.

Depoldering, flood bypasses, controlled floodable areas, onsite water retention, programmed debris deposition, superdune, superdyke are the key actions for resistance to and/or retention of the environmental dynamics at the basis of the newly developed visions and interventions.

Depoldering

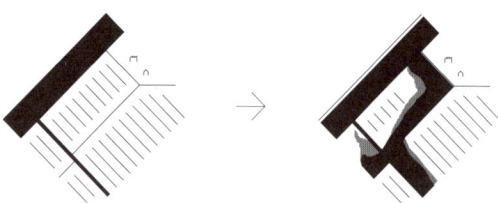

Depoldering (or de-reclamation) is an alternative to increasing the defences. It implies restoring original water dynamics (sea water, river water and rainwater) to some or entire parts of the territory formerly drained through the construction of dykes against incursions by external waters, of adequate drainage networks and the installing of pumps for letting out rainwater and seepage water from the dammed area (MTPWWM, 2006; Desvigne, 2008; De Koning and van der Meulen, 2009; Slabbers and Koole, 2009; Waterwegen en Zeekanaal NV Afdeling Zeeschelde, 2012).

The depoldering consists in the partial or total shutdown of the draining pumps, the breakage or removal of embankments in one or more points of the system, and any eventual excavation and guiding actions to trace preferential directions for the water. The area goes back to being subjected to permanent or periodic flooding due to rainwater, seepage water, and/or the water from the river or the sea. The water progressively invades the area starting from the lower parts.

The depoldering has a different effect when realized along the river rather than along the seacoast. In the first case, the removal of all or part of the river dyke leads to the reconnection of the watercourse to the territory. The riverbed and/or its floodplain expand. When the levels of the watercourse rise, the river water can spread out transversally from the direction of the river's flow. The expansion area allows the river's levels at normal flow levels and periods of fullness to be lowered. This reduces the risk of flooding. In the second case, along the seacoast, removing all or part of the sea defence embankments restores the dynamics of the tides to the land environments. The new wetlands (such as lagoons and marshes) serve as a filter between the land (and freshwater) conditions and sea conditions (salt water). This reduces the overall pressure of the sea on the land.

In any case, depoldering carries to greater expanses of wetlands in the deltaic landscape, which are opportunities to increase areas of high ecological value and/or for recreation. However, depoldering requires rethinking of land use. Beyond the areas that permanently become water areas, those that remain dry are susceptible to flooding. This means some infrastructure and settlements must be secured if not removed to be possibly located elsewhere, the system of connections (roads, bridges, pedestrian and cycle paths) must be reorganized, and the crops must be reconsidered in favour of crops adapted to periodic invasions of water.

Flood bypass

the orography and existing activities and land uses. The terrain is then modelled by tracing out the preferential water routes and the dikes of the new river branch. Downstream, at the point of re-entry a second cut is done on the main river embankment and, possibly, a manufactured hydraulic control is installed. From the diversion point, the riverwater finds an alternative route which it invades from the lower parts. The water entering the bypass re-joins the river further downstream through the re-entry point. The bypass expands the river bed in a strategic point of the territory (lower parts, less inhabited areas etc.). The river water finds a new regulated space in which to expand. This lowers the river's flood levels, and in some cases also its normal flow levels. The risk of the watercourse overflowing decreases. The opening of a river bypass results in the extension of the wetlands and in the general increase of the areas for 're-naturing'. The areas of the bypass are also suitable for accommodating different recreational practices.

The realisation of a river bypass implies the reorganization of the land use for the area involved. Some infrastructure and settlements will have to be secured if not removed to be possibly located elsewhere, the system of connections (roads, bridges, pedestrian and cycle paths) reorganized, crops adapted to inundations can be planted in the parts of the bypass that are periodically flooded.

Controlled floodable area

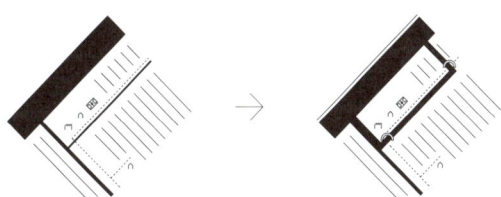
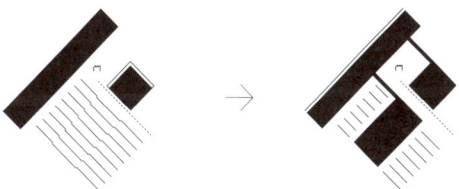

The flood bypass (also known as flood channel) is an operation aimed at extending the riverbed and/or its floodplain through the opening of a new branch of the river (MTPWWM, 2006).
A stretch of the existing embankment is demolished. There, a water regulation system is installed. The bypass route goes downstream taking into account

The area for controlled flooding (or overflow basin) is a device which increases the river's alluvial plain area (MTPWWM, 2006; Waterwegen en Zeekanaal NV Afdeling Zeeschelde, 2012).
A hydraulic device for regulating the river's flow is placed on the riverbank to connect the river to the controlled flood area. If the hydraulic intake is

not directly linked to the controlled flooding area, a diversion channel connects the river to the depressed area for flooding. Sometimes the area to be flooded is confined by levees enclosing the basin. Other times the area for controlled flooding is already a body of water (lake, pond, etc.) whose banks offer some margin of water fluctuation.

The controlled floodable area works especially for the case of flood conditions. Part of the river's discharge can indeed be diverted towards the controlled flooding area through the opening on the riverbank. The water can expand until it reaches the established safety limits. In this way the river's flood levels are lowered. The watercourse's risk of overflowing decreases.

The realisation of an area for controlled flooding in many cases allows to maintain the existing land uses. Other times, the land use is changed, allocating the overflow basin as a wetland to increase biodiversity and/ or a park to provide recreational opportunities. Still other times the controlled flooding area has a further storage capacity (seasonal storage), which results in water storage for irrigation, for example.

The making of a controlled floodable area requires various interventions depending on the frequency with which the area is subjected to flooding and the flexibility with which the existing land uses can be adapted to that flooding. Some infrastructure and settlements must be secured or removed and be located elsewhere, the system of connections (roads, bridges, pedestrian and cycle paths) reorganized, the farming layout rethought in favour of agricultural crops adapted to periodic water invasion or that otherwise do not require burdensome agrarian arrangements for avoiding structural damage in the event of flooding.

Onsite water retention

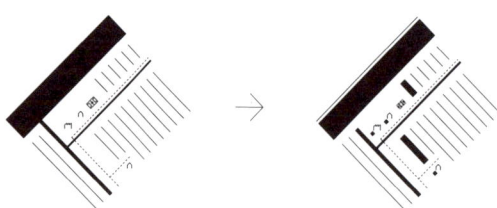

Onsite water retention is an operation aimed at reducing pressure on water networks (Tjallingii, 1996; De Greef, 2005; Ranzato, 2011).

The forms of onsite water retention are numerous and vary also depending on the scale at which they operate: tanks, ponds, depressed areas, ditches, etc. Rainwater and/or sewage waters are either retained locally to be released slowly or possibly treated and stored for reuse.

Retaining the water locally results in the reduction of the discharge released into the drainage system and eventually into the river as the final destination, to lessen the chances of flooding. For the depressed areas this may also mean reducing the pumping and the consequent energy consumption. Furthermore, the possible reuse of retained water reduces the pressure on the supply system and thus on the river as a source of fresh water. The consequent increased availability of fresh water in the river and the storage of fresh water at the local level can be crucial to countering saltwater intrusion.

The devices needed for onsite water retention offer the possibility to rearrange spatially some parts of the territory. The new spaces for water can be an opportunity to locally extend the areas for recreational parks and resting places and to locally increase biodiversity. Onsite water retention involves the reorganization - even only partially - of the water networks. It must be taken into account that the space required for storing (and possibly treating) the water must be subtracted from other activities and land uses unless ingenious integrations are made.

Programmed debris deposition

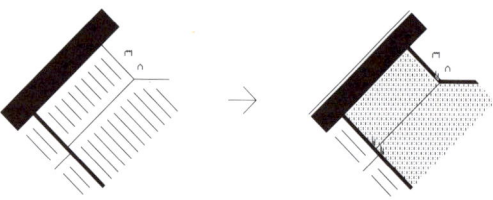

Programmed debris deposition (or filling) is a reclamation operation that takes advantage of river dynamics and is based on the principle of *building with nature* (LSU Coastal Sustainability Studio, 2011; Latitude Platform, 2012). The programmed debris deposition consists in: the construction of one or more diversion channels (*fillers*) to divert a certain proportion of the flood's flow towards depressed areas or areas at any rate lower than the water level of the diversion channels, the enclosing of the depressed areas to be reclaimed by levees that surround the basin (*reclaimed land*), and the construction of canals that allow to drain the waters of the reclaimed land. The diversion channels pour the dirty water (with high sediment content) onto the land to be reclaimed. By stagnating, the water deposits the sediments on the bottom. Once decanted of its sediments, the excess water is forced out and conveyed into the appropriate drainage channels. The sediment deposits raise the level of the land.

The programmed debris deposition allows the land area to be extended into the watery area. This operation can be used to raise the levels of depressed land, both reclaimed and non. In this way, for the area the possibility of water infiltration from outside bodies diminishes, as does the use of draining pumps and the risk of flooding in general. Debris deposition, by taking advantage of the river's flooding, moderates the risk of flooding.

Debris deposition is a process and as such puts the involved basins 'on hold', making them difficult to use for other purposes. Indeed, during the filling operation, the basin for debris deposition is subject to periodic flooding. Therefore, the use of its space must be reorganized and activities incompatible with flooding must be suspended. Often infrastructure (roads, bridges, pedestrian and cycle paths) and settlements must be secured. In some cases their removal and possibly their relocation become necessary.

Superdune

The superdune is a system aimed at strengthening sea defences, based on the principle of *building with nature*. Superdune entails exploiting marine currents and coastal winds for the construction or reinforcement of sand dunes along the coastline (Berger, 2009).

The realisation of the superdune requires two stages: in the first stage, sand from the sea bottom is pumped to build the superdune in an area adjacent to the coast and in a strategic position with respect to the coastal currents and winds; in the second phase, currents and winds distribute sand from the artificial sand dune along the coast.

The superdune is a device that strengthens the sand dunes along the coastline. As a transitional space between land and sea, the sand dunes protect inland areas from winds, storm surges, sea level rise and, in some cases, saltwater intrusion. The superdune is a space in transformation, slowly consumed in order for the coastline to receive sedimentation. However, during the transformation process, the superdune can serve as a potential support for the enhancement of biodiversity and recreational practices. At the end of the process of the superdune's 'dissolution' by currents and winds, the coastal dune ecosystem is reinforced and becomes even more suitable to accommodate biodiversity and recreational practices. In addition, the strengthened dune system will more effectively serve as a windbreaker.

The realisation of a superdune implies also some minuses. Indeed, it requires large amounts of sand. The removal of sand from the open sea asks for considerable amounts of energy for pumping. Moreover, the extraction of sand alters the seabed currents, the effects of which are difficult to predict.

Superdyke

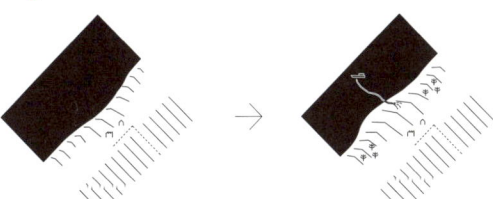 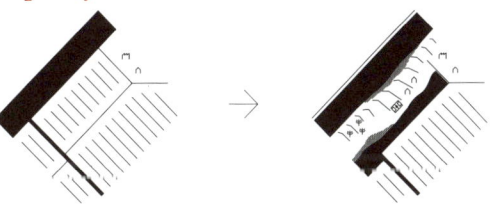

The superdyke is a device for securing land areas from sea and river dynamics typical of deltas and estuaries (Sijmons, 2002; MTPWWM, 2006). Large amounts of land are accumulated longitudinally to the river's flow, sometimes along the trace of existing levees. The superdyke may have a contained, linear extension or may surround an area to isolate it (partially) from external systems. The cross section of a superdyke can vary considerably, but is always quite extensive.

The superdyke's large size makes it a strong resistance device, a barrier difficult to penetrate for the river or sea waters and, under certain conditions, even for the subterranean currents (groundwater). The superdyke is a reformulation of the traditional defence embankment. In addition to its large size, the superdyke is also characterized by its pronounced multifunctionality. Regardless of whether it protects an area from external agents, the superdyke itself is a 'safe space to live'. This means it can accommodate a variety of activities that elsewhere would be impossible or unsafe. In addition, along its length, the host activities can vary in relation to the gradient of water levels on its slopes. In fact, the closer one gets to the mass of the superdyke, the possibility of direct contact with the dynamics of water increases. The superdyke can give room to diverse infrastructures, such as railway lines, roads, bicycle and pedestrian paths, residential and productive buildings or areas, wetlands, forests, orchards, cultivated fields, etc. It can also serve as a windbreaker.

The superdyke is a device that requires large amounts of land in order to be built. The massive use of land can be considered as rational occurring in its vicinity, such as excavation works, necessary to widen or deepen the river bed and/or expand its floodplain.

References

- Berger, A., 2009. Systematic design can change the world. Zeist: SUN Publishers.
- De Greef, P., 2005. Rotterdam waterstad 2035. Rotterdam: NAI.
- De Koning, R., van der Meulen, Y., 2009. Depoldering Noordwaard. In: Dutch Dialogues, Meyer, H., Morris, D., Waggonner, D. (eds). Amsterdam: Sun.
- Desvigne, M., 2008. Intermediate natures: the landscapes of Michel Desvigne. Berlin: Birkhauser.
- Ebert, S., Hulea, O., Strobe D., 2009. Floodplain restoration along the Danube: a climate change adaptation case study. In: Lessons for climate change adaptation from better management of rivers, Pittock, J. (ed). London: Earthscan.
- Latitude Platform, 2012. Veneto 2100: Living with water. In: Making City: catalogue of the International Architecture Biennale Rotterdam, Brugmans, G., Petersen, J., W. (eds). Rotterdam: IABR.
- LSU Coastal Sustainability Studio, 2011. Blue ribbon resilient communities: envisioning the future of America's Energy coast. Ppt presentation at America's energy coast leadership forum, Plaquemines Parish, May 16, 2011.
- McHarg, I., 1967. Design with nature. New York: Natural History Press.
- MTPWWM (Ministry of Transport, Public Works and Water Management), 2006. Spatial Planning Key Decision 'Room for the River'. The Netherlands.
- van Leuwen, C., G., 1973. Ekologie. Faculty of Architecture, Delft University of Technology.
- van Wirdum, G., 1982. The ecohydrological approach to nature protection. In: Research Institute for Nature Management, Annual Report 1981, Arnhem, Leersum, Textel, 60-74.
- Slabbers, S., Koole, S., 2009. River widening: Overdiepse Polder. In: Dutch Dialogues, Meyer, H., Morris, D., Waggonner, D. (eds). Amsterdam: Sun.

- Ranzato, M., 2011. Integrated water design for a decentralized urban landscape. Thesis (Ph.D.). University of Trento.
- Sijmons, D., 2002. =landscape. Amsterdam: Architectura & Natura Press.
- Tjallingii, S., P., 1996. Ecological conditions. strategies and structures in environmental planning. Wageningen: Institute for Forestry and Nature Research. Thesis (Ph.D.). Delft University of Technology.
- Waterwegen en Zeekanaal NV Afdeling Zeeschelde, 2012. Sigmaplan Dedwige-Prosper-projec. In: Ontmoet the Schelde. Antwerpen: Agentschap voor Natuur en Bos.

Published by
LISt Lab Laboratorio Internazionale Editoriale
Italy - Via Esterle, 26
38122, Trento
Spain - Netherlands
info@listlab.eu
www.listlab.eu

Production
GreenTrenDesign Factory
Piazza Manifattura, 1
38068 Rovereto (TN) - ITALY
T: +39 0464 443427
info@greentrendesign.it

Author
Maria Chiara Tosi

Editorial Director
Pino Scaglione

Art Director
Massimiliano Scaglione

Editorial Assistant
Gioia Marana

Graphic Design
List Lab e/and Marc Sánchez

Translations
Just!Venice

Scientific Board of the LISt Edition
Pepe Barbieri, Rosario Pavia (Università di Chieti) Eve Blau (Harvard GSD), Maurizio Carta (Università di Palermo), Eva Castro (Architectural Association London) Alberto Clementi (Università di Chieti), Alberto Cecchetto (Università di Venezia), Stefano De Martino (Università di Innsbruck), Corrado Diamantini (Università di Trento), Antonio De Rossi (Università di Torino), Franco Farinelli (Università di Bologna), Carlo Gasparrini (Università di Napoli), Manuel Gausa (Università di Barcellona/Genova), Giovanni Maciocco (Università di Sassari/Alghero), Antonio Paris (Università di Roma), Vanni Pasca (Università di Palermo), Mosè Ricci (Università di Genova), Roger Riewe (Università di Graz), Pino Scaglione (Università di Trento).

Atlas Drawings Matteo Carli, Alberto Oss Pegoraro, Marta de Marchi, Andra-Loredana Medeleanu, Sorin Tudor Bompa, and students participating in the Erasmus IP 2009-2011 "Summer school on the Po river delta. Creating scenarios in fragile territories"

All rights reserved
© of the edition, LISt Lab
© of the text, the authors
© of the images, the authors; the authors recognize possible rights for the published images.
Satellitary Images (Landsat 7)
Credit: courtesy of the U.S. Geological Survey, Department of the Interior/USGS
Pictures, Stefano Munarin
Printed and bound in the European Union, July 2013

Printing
Rubbettino print

ISBN 9788895623870

Promotion in Italy and International distribution
Messaggerie, Milano; Actar D, New York.

LISt Lab is an editorial workshop, based in Europe, that works on the contemporary issues. LISt Lab not only publishes, but also researches, proposes, promotes, produces, creates networks.

LISt Lab is a green company committed to respect the environment. Paper, ink, glues and all processings come from short supply chains and aim at limiting pollution. The print run of books and magazines is based on consumption patterns, thus preventing waste of paper and surpluses. LISt Lab tends to raise the authors and market's awareness of a new editorial culture based on smart resource management.